Applied Veterinary Andrology and Frozen Semen Technology

nipa® books for everyone
PAPERBACKS

NIPA® GENX ELECTRONIC RESOURCES & SOLUTIONS P. LTD.
New Delhi-110 034

Applied Veterinary Andrology and Frozen Semen Technology

by :

M.K. SHUKLA

Assistant Professor/ Scientist
Department of Animal Reproduction, Gynaecology and Obstetrics
College of Veterinary Science & Animal Husbandry (MPPCVV)
Kuthulia, Rewa- 486001, Madhya Pradesh

nipa books for everyone
PAPERBACKS

NIPA® GENX ELECTRONIC RESOURCES & SOLUTIONS P. LTD.

New Delhi-110 034

books for everyone
PAPERBACKS

NIPA® GENX ELECTRONIC RESOURCES & SOLUTIONS P. LTD.

101,103, Vikas Surya Plaza, CU Block
L.S.C. Market, Pitam Pura, New Delhi-110 034
Ph : +91 11 27341616, 27341717, 27341718
E-mail: newindiapublishingagency@gmail.com
www: www.nipabooks.com

For customer assistance, please contact
Phone: + 91-11-27 34 17 17
Fax: + 91-11- 27 34 16 16
E-Mail: feedbacks@nipabooks.com

ISBN:978-81-19002-21-4

Composed and Designed by NIPA®.

Where emotions are involved,
Words cease to mean.
There are no words to express my regards to
My Mummy & Papa
Your endless love and shelter over me
Incites me to achieve new heights in life
Words can do no better than abating
Your enormous blessings and sacrifices for me.

Dedicated to....

My Beloved Parents

Preface

To the almighty God, who rules all things well,
I give all the glory

Ample literature covering various aspects of veterinary andrology and frozen semen technology is available but need for a book incorporating practical aspects of the subject was always felt for the students of andrology and semen biology and scientists working in the area. This book is aimed to fill this void in literature by providing insight into various applied aspects of veterinary andrology, frozen semen technology and artificial insemination with the help of relevant illustrations based on author's experience and research in the subject. Theoretical aspects of the subject have been deliberately omitted as ample literature on the topic has already been published.

This book has been written to supplement the requirements of the scientists and Managers working in frozen semen production station, Semen Quality Control Laboratories, Andrological diagnosis Laboratories and students of Andrology and Artificial Insemination. It incorporates topics mentioned in syllabus for Practical course of Andrology and Artificial Insemination of undergraduate students of Veterinary Science. This will also be helpful to the graduate students of Animal Reproduction or Veterinary Andrology as a teaching aid.

This book is based on author's experience and research in the subject during his postgraduate dissertation work and also during his association with prestigious organizations like G.B. Pant University of Agriculture & Technology (GBPUA&T), Animal Breeding Research Organization (ABRO), National Dairy Development Board (NDDB) and Sardar Krushinagar Dantiwada Agriculture University (SDAU). The author feels deeply obliged to record his heartfelt sense of gratitude and indebtedness to all the mentioned organization for providing ultra modern laboratories and state of art facilities which enabled author to learn fine aspects of Andrology and Frozen semen technology and to carry out some remarkable work which lead to the rare photographs presented in the book. The author feels fortunate to be associated with eminent scientist like Prof. A.K. Misra as teacher and guide, during his postgraduate degree programme. I fumble for words to venerate my heartfelt sense of gratitude to him for initiating my interest in the subject and for his meticulous guidance, constant motivation and constructive criticism received from time to time.

It is like a drop in the ocean of words that can never reach it's mark to acknowledge understanding and constant encouragement of my wife Rashmi and daughter Hashya. Most of the time spent for preparation of this manuscript was derived from their share which they accepted without complaints. Though all possible efforts have been made to omit errors but to err is human. Therefore, possibilities of errors exist in the manuscript. The readers are requested to report the same to the author and your valuable suggestions for further improvement of the manuscript are also welcome.

M.K.Shukla
shuklamanishis@gmail.com
Mobile : 9753385220

Contents

Section-B : Semen Processing & Ccyopreservation

Section-C : Semen Evaluation

Section-D : Artificial Insemination

Section A

Andrology

Chapter 1

Orgena Genitalia Masculina (Male Reproductive Organs) in Domestic Animals

Knowledge of male reproductive system and its physiological functions is essential for understanding male reproductive behavior and for effective reproductive management of breeding bulls. Therefore, personnel involved in management of breeding bulls, semen collection, handling and artificial insemination must have sound knowledge of male reproductive anatomy and physiology of breeding bulls. This section provides a recapitulation of the basic physiology and anatomy of male genitalia.

Functions of Male Genital System

(a) Primary function of the male genital system is to monitor reproduction by providing cutaneous and glandular system of male gonads

(b) Secondary function of male genital organs is to provide glandular, ductular, copulatory and auxillary system associated with male reproduction.

Based on the above functions the male reproductive system can be divided into the following components:

1.1 Primary Andrological organs
  1.1.1 Scrotum
  1.1.2 Testis

1.2 Secondary Andrological organ
  1.2.1 Excurrent ducts
    1.2.1.1 Ductus epididymis
      Caput
      Corpus
      Cauda
    1.2.1.2 Ductus deferens and Ductus ampulla
    1.2.1.3 Ductus urethra
  1.2.2 Accessory sex glands
    1.2.2.1 Seminal Vesicles/ Vesicular glands
    1.2.2.2 Prostate
    1.2.2.3 Bulbourethral gland/ Cowpers glands
  1.2.3 Copulatory organs
    1.2.3.1 Prepuce
    1.2.3.2 Penis

1.1 The Primary Andrological Organs

1.1.1 Scrotum

This is an oblong pendulous cutaneous pouch which encloses the testis. It is an unisacular structure separated by a projection of muscular septum internally and a scrotal ridge externally. It is located between the thighs except boar and cats where it is located caudal to thighs and caudal and ventral to the ischiatic arch. It is composed externally of skin and bears scanty hairs except cat, sheep and goat. Beneath the skin, tunica dartos (consisting of fibroelastic tissue and unstrapped muscle) is present which is connected closely with tunica vaginalis and gubernaculums forming the scrotal ligament. The scrotum has a medial septum formed by the dartos dividing the scrotum in to two halves. The spermatic fascia/ scrotal fascia is fastened by loose areolar tissue to the dartos and this fascia is lined by the tunica vaginalis communis. It is thin above but thicker in the scrotum. Tunica vaginalis communis is the superficial layer and is an extension of peritoneum passing through the abdominal wall at the inguinal canal. It corresponds to the parietal peritoneum of the abdominal cavity. Tunica vaginalis propria is the deeper layer which corresponds the visceral layer of the peritoneum of the abdominal cavity. Tunica albuginea layer is a tough layer composed of fibromuscular tissue present beneath the visceral layer of the vaginal process.

Functions

(i) It encloses the male gonad and provide suspensory attachment of testis with the animals body.

(ii) It envelops the epididymis and certain contents of the distal portion of the spermatic cord.

(iii) It maintains the temperature of the testes through thermoregulatory mechanism.

1.1.2 Testis

Testis, the male gonads, are paired and are suspended in the scrotum. Consistency of the testis is usually turgid; parenchyma is yellowish to reddish in color and bulges on section. The testis is suspended by a delicate double layer of peritoneum connecting visceral and parietal layer of the vaginal process (Tunica vaginalis communis and Tunica vaginalis propria). The mesoorchium continues upto dorsolateral abdominal wall. Testis is supported by one of the two scrotal pouches where it is held by its tunics and by spermatic cord which is composed of-

- Spermatic artery tortuously coiled just dorsal to the testes.
- Spermatic vein- It forms a plexus of veins, the pampiniform plexus around the spermatic artery.
- Internal cremester muscle
- Lymaphatic vessels
- Autonomic nerves from the renal and posterior mesenteric plexus which forms the spermatic plexus around the vessels in the cord.
- The vas deferens
- Tunica vaginalis propria

Spermatic cord is comparatively longer in boar.

Distal end of the testis is attached to the scrotum by scrotal ligament. Tunica vaginalis propria covers the testis (thin serous membrane). Beneath the tunica vaginalis propria is tunica albuginea (thick connective tissue capsule). Septa from tunica albuginea penetrate the testicular parenchyma to join at the mediattinum testis and forms lobules of the testis with exception of the horse. Seminiferous tubules are highly coiled tubules present in these lobules and they account for about 50 % of the testicular mass. The seminiferous tubules lined by germinal epithelium produces spermatozoa.

Length of the Seminiferous Tubule in Different Animal Species

Boar	6000 m
Bulls	5000 m
Ram	4000 m
Dog	150 m
Cat	25 m

Between seminiferous tubules Islands of Interstitial/ Leydig cells are present. Seminiferous tubules converge at the apex of the lobule at the receptacle (Marin-Padilla) to join the straight tubule or tubuli recti which enters the rete testis. Rete testis is a structure of anastomosing spaces located in the mediastinum testis. Mediastinum testis is absent in stallion and collecting tubules join the efferent tubules. Sperms pass from rete testis to the head of the epididymis through 6 to 24 efferent tubules or ducts. The transportation of sperm cells till this phase is due to presence of testicular fluid.

1.2 Secondary Andrological Organ

1.2.1 Excurrent Ducts

1.2.1.1 Ductus Epididymis

It is composed of a single, tortuous, coiled tubule. It has a smooth muscle coat that moves sperm towards the vas deferens in peristaltic wave. The epididymis is closely attached to the surface of the testis proper by fibrous tissue. It consist of three parts:

1. Caput Epididymis/ Head of the epididymis - Proximal / spermatic cord end of the epididymis. It is more or less flat, broad and V shaped.

2. Corpus epididymis/ Body of the epididymis- It is intermidiate part. It is narrow.
3. Cauda epididymis/ tail of the epididymis – The distal end of the epididymis is enlarged and serves as a storage site for spermatozoa.

Length of the Epididymis in Different Animals

Bulls - 30 m (100 feets)
Boar - 50 m (155 feet)
Horse - 20 m (65 feet)

Functions of Epididymis

1. Absorption – The epithelial cells of the epididymis, especially of the tail region absorb the excess fluid released by the testis and helps in concentration of spermatozoa.
2. Secretion- Various secretions of the epididymis provides nutrition and this helps in maintaining the viability of the spermatozoa.
3. Maturation- Sperm cells undergo various changes amounting to maturation during storage in epididymis e.g. migration of proximal protoplasmic droplet from neck to the distal end of the mid piece in bulls. This is essential for increasing the motility and fertilizing ability of the bulls.
4. Transportation- The transportation of spermatozoa from head of the epididymis till ejaculation is mainly due to action of ciliated epithelium and action of peristaltic waves of the muscle fibers in the duct. The duration of transport of spermatozoa in different species of domestic animals is as follows:

   Bulls 10 days
   Ram 13-15 days
   Boar 9-12 days
   Stallion 8-11 days

5. Storage of spermatozoa- The epididymis also serve to store the spermatozoa. They can store upto 3-4 days production of spermatozoa by the testis (upto 75 X 10^9). Cauda epididymis is responsible for nearly 50 % storage by the extra gonadal sperms. The sperms, however, remains metabolically inactive during storage in the epididymis.

Position of epididymis in respect to testis in different domestic animals

Species	Caput	Corpus	Cauda
Bull	Dorsal	Caudal and medial to testis	Ventral
Ram & Buck		Similar to bulls	
Stallion	Cranial	Dorsal to testis	Caudal and loosely attached to the testis
Boar	Ventral	Cranial	Dorsal
Dog	Cranial	Dorsal	Caudal

Testicular biometry in domestic animals

Species	Length (cm)	Diameter (cm)	Weight (g)
Stallion	7.5 – 12.5	4-7 (Dorsal-ventrally) 5 (thickness)	200 - 300
Bull	10 – 15	5 - 8.5	200 - 500
Ram & Buck	7.5 – 11.5	3.8 – 6.8	200 - 400
Boar	10 – 15	5 – 9	500 - 800
Dog	2 – 4	1.2 – 2.5	7 - 15
Cat	1.2 - 2	0.7 – 1.5	-

1.2.1.2 Ductus Deferens / Vas deferens

Ductus deferens extends from the tail of the epididymis to the colliculus seminalis of the pelvic urethra. Walls of the vas deferens are thick and has got uniform thickness from its origin to the dorsal surface of the bladder (3mm - bulls and 6 mm - stallion) till the fusiform terminal glandular part called ampulla. The lumen of the vas deferens is quit small. The ampulla is absent in dogs and cats and are very small in boar. In bulls ampullae lie dorsal to the neck of the bladder parallel to each other and in close apposition. Caudally they narrow and pass under the body of the prostate and opens into the cranial portion of the pelvic urethra through a rounded prominence called colliculus seminalis. Ampullae lie parallel to the body of the epididymis and medial and cranial to it. In stallion they lie dorso-medial to the body of the epididymis and testis. In boar ampullae are located cranial and medial to the testis.

1.2.1.3 Ductus Urethra

Urethra in males is a common passage for the urine excretion and the passage of the semen. It has three parts-

a. Pelvic part- In bulls it is about 20 cm in length and is situated on the pelvic floor. The pelvic urethra is enclosed by heavy urethral muscles. Plevic urethra of bulls, ram and boar contains many glands similar to those in prostate. This diffuse layer of glands is called the disseminate part of the prostate and not the urethral glands. Urethral glands are prominent in man.

b. Bulb of urethra- It is the extra pelvic part situated at the ischial arch and bends ventral to the pelvis.

c. Penile urethra- It runs inside the penis proper.

A remnant of paramesonephric duct, situated on the caudal dorsal surface of the bladder between the ampullae of the vas deferens and the seminal vesicles and cranial to the prostate, is called uterus musculinus.

Muscles of the Pelvic Urethra

1. Urethral muscle- This is circular striated muscle surrounding the pelvic urethra. It aids to ejaculation and micturation by it's forcible contractions.

2. Bulbocavernosus muscle- This is striated circular muscle surrounding penile urethra. It is thickest at the root of the penis. It functions to empty the extrapelvic urethra. It's rhythmic contractions compress the corpus spongiosus proximally and assist in erection.

1.2.2 Accessory Sex Glands

The following accessory male genital glands are present in male animals with certain exceptions:

1.2.2.1 Seminal Vesicles/ Glandulae Vesiculares/ Vesicular Glands

Seminal vesicles are paired glands with distinct lobulations in bovine. They are absent in dogs and cats and are smooth in equines. They are proportionately very large in boar and cover the caudal portion of the bladder and extends into the abdominal cavity. A portion of the seminal vesicles is covered with peritoneum in bulls, stallion and boar. Seminal vesicles are located on the floor of pelvis, cranial and lateral to the ampullae and neck of the bladder. They open in pelvic urethra in close proximity to the opening of vasa deferentia. The lobes of the vasicular glands have small central dilatation in equines. This structure bears gland like nature in most of the animals and hence the name vesicular gland, except equines where it is more bladder like with the glands located on the wall.

Biometry of seminal vesicles in different domestic animals

Species	Shape	Length (cm)	Breadth (cm)	Thickness (cm)	Diameter (cm)	Weight (g)
Bull	Lobulated	10-15	2-4	2	2-4	75
Stallion	Smooth	15-20	2.5-5	5	2.5-5	-
Boar	Large	12-15	5-8	4	5-8	200
Ram	Lobulated	4-5	2	1.5	-	5

Functions of Vesicular Glands

It secretes a clear fluid and adds volume, nutrients and buffer to the semen which may form up to 50 % of the total semen ejaculate. They do not store spermatozoa.

Characteristics of Vesicular Secretions

- Vesicular secretions contain protein, fructose ascorbic acid, citric acid, potassium bicarbonate, acid soluble phosphate and several enzymes. They are more alkaline as compared to the secretions of the prostate.
- Most of the seminal fructose in mammals comes from the secretions of the seminal vesicles.
- In stallion seminal secretions contributes gel to the ejaculate.
- In boar the secretions are milky and highly viscous and contain high inositol content and it also contains ergothionine.
- In bulls secretions are yellow due to high riboflavin content.

1.2.2.2 Prostate Gland

Prostate gland is located on the floor of the pelvis at the cranial end of the pelvic urethra caudal to the neck of the bladder. It opens into pelvic urethra lateral to the colliculus seminalis by means of multiple ducts. There are only two excretory ducts in dogs. The characteristics of the prostate gland in different animal species are as under:

Bull

Prostate gland in bulls consists of two parts:

- Pars propria (body of the prostate)
- Pars disseminate

The width and thickness of the Pars propria (body of the prostate) is 2.5-4 and 1-1.5 cm, respectively. Pars disseminate surrounds the pelvic urethra dorsally it is 1-1.5 cm thick and 10-12 cm long and is covered by urethral muscles. Prostate gland can be felt per rectum as a distinct transverse oval protuberance on the cranial end of the pelvic urethra.

Ram

Body of the prostate is absent. Prostate gland is diffused over a large portion of the pelvic urethra.

Stallion

In stallion the prostate gland consists of two lateral lobes connected by an isthmus. It is located over the neck of the bladder and cranial portion of the urethra. Length and width of the gland is 5 and 2.5 cm, respectively.

Boar

Body of the prostate gland is located dorsal to the bladder and is covered by the seminal vesicles. The disseminate part of the prostate gland around the pelvic urethra is quit extensive as in bulls and rams.

Dogs

Prostate gland is relatively large in dogs and may be quit enlarged in older dogs. They are located at the cranial border of the pubis. It surrounds the neck of the bladder and urethra at their junction and is thin dorsally. The dimensions of the prostate may vary from 1.7 X 2.6 X 0.8 cm to a nearly spherical with diameter of 2 cm.

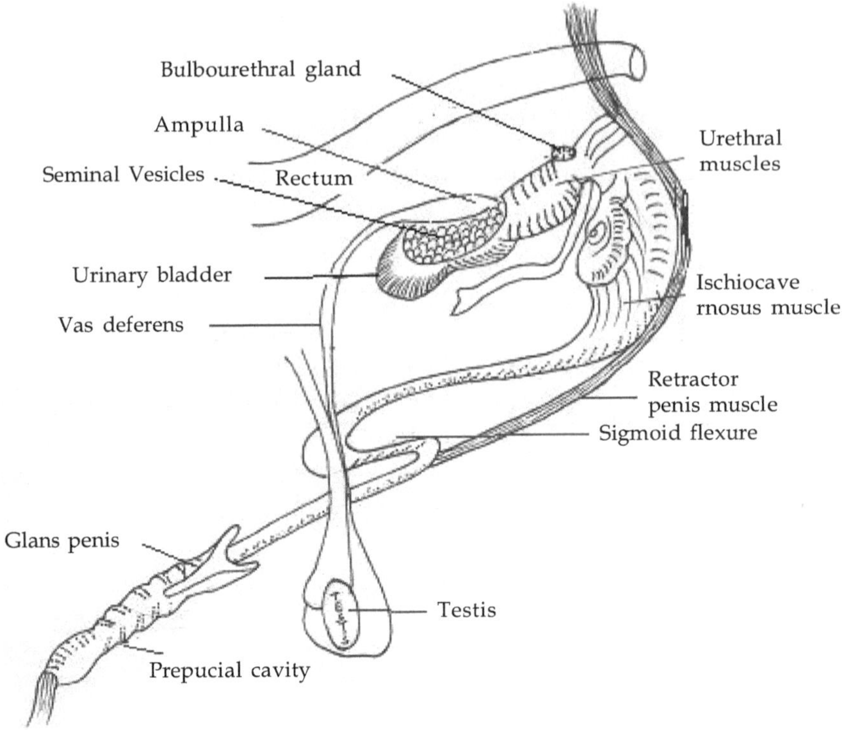

Male Reproductive Organs - Bull

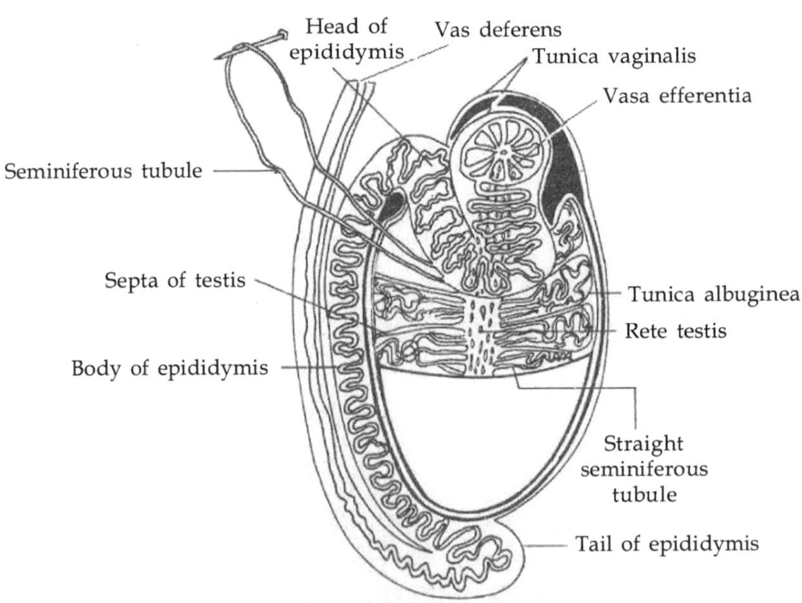

Schematic diagram depicting the tubular system of testis

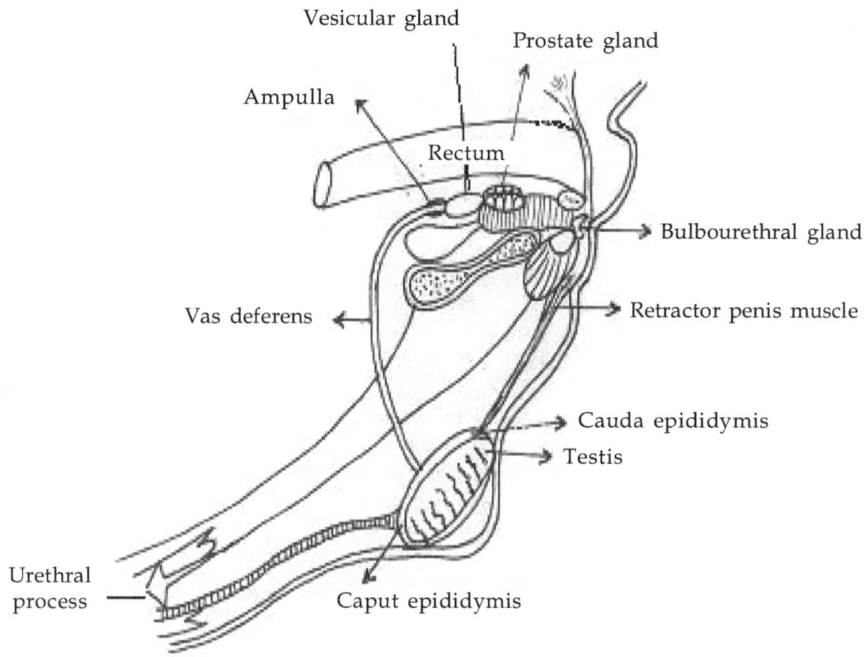

Male Reproductive Tract - Stallion

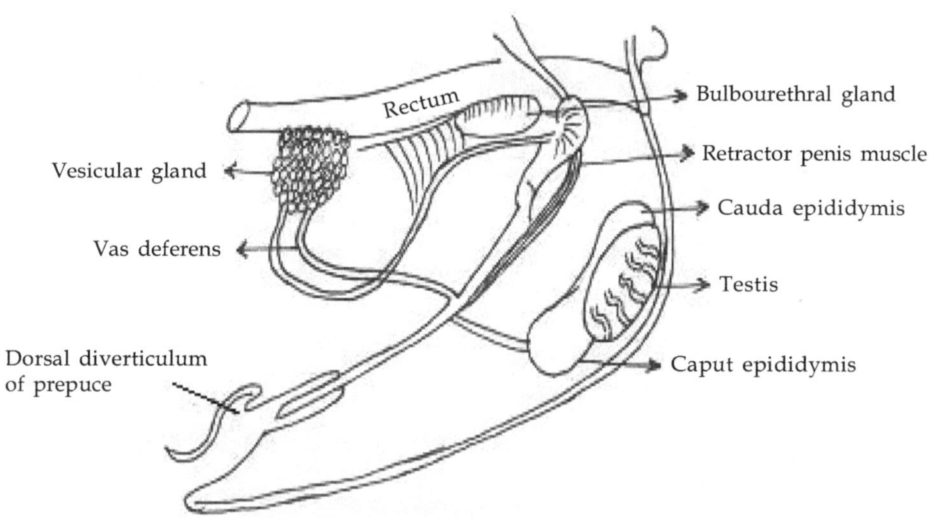

Male Reproductive Tract - Boar

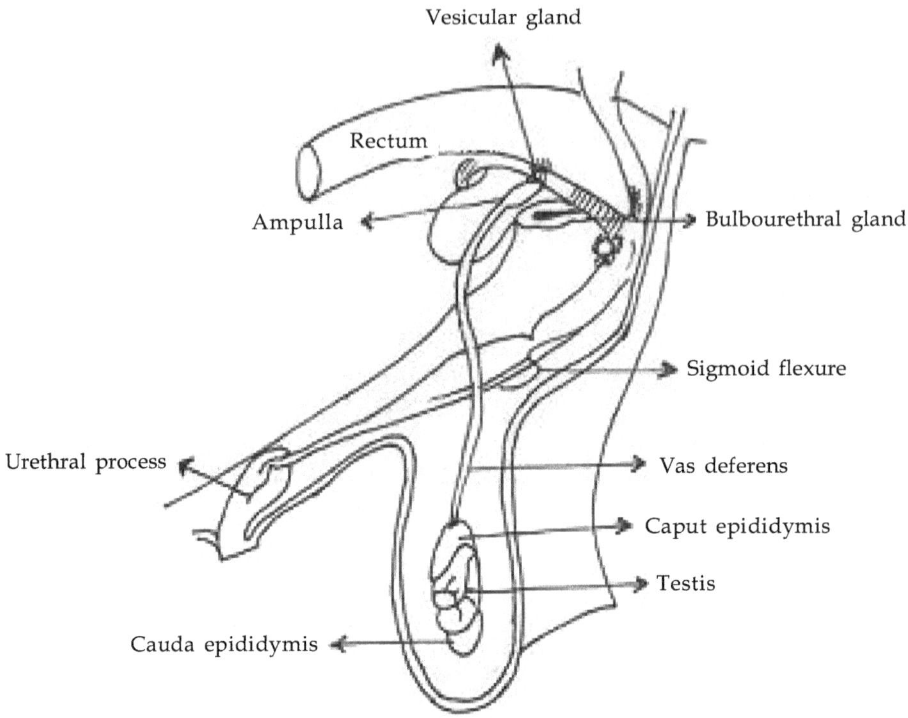

Male Reproductive Tract - Ram

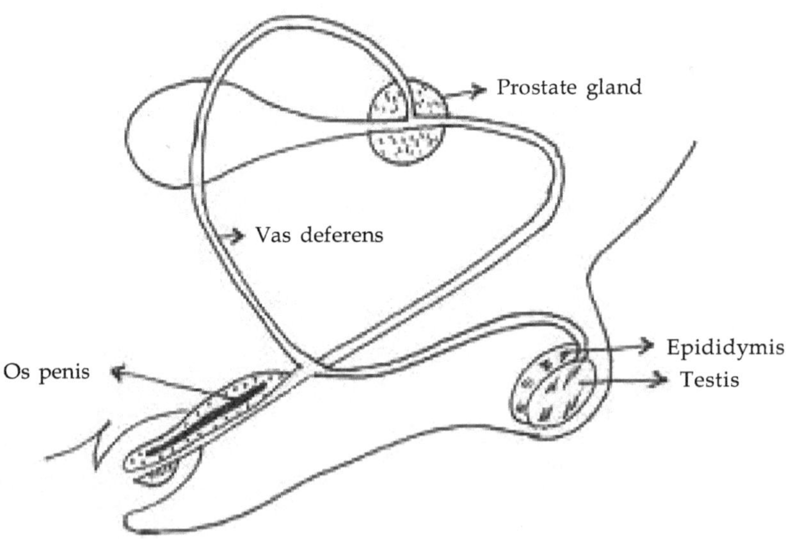

Male Reproductive Tract - Dog

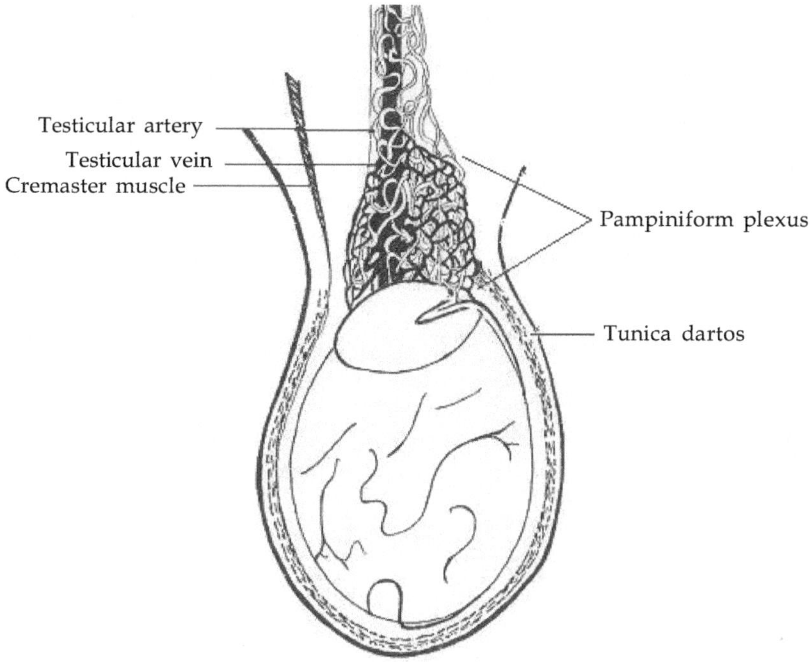

Schematic diagram depicting vascular system and thermoregulatory apparatus of the testis

Persistent penile frenulum

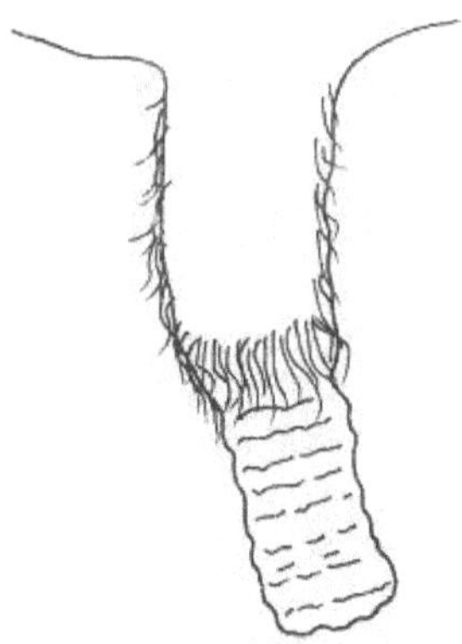

Eversion of prepuce

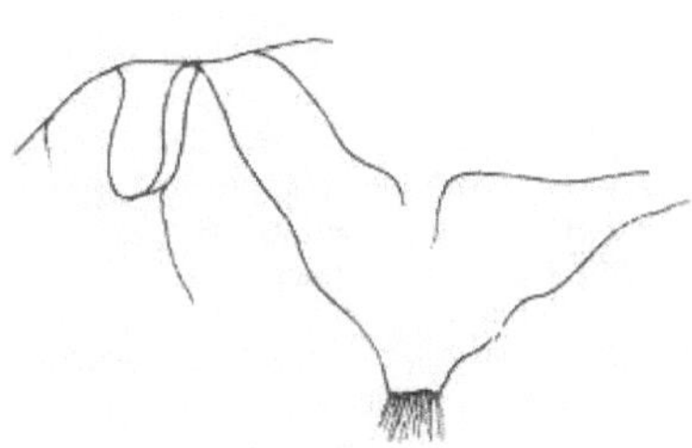

Pendulous sheath

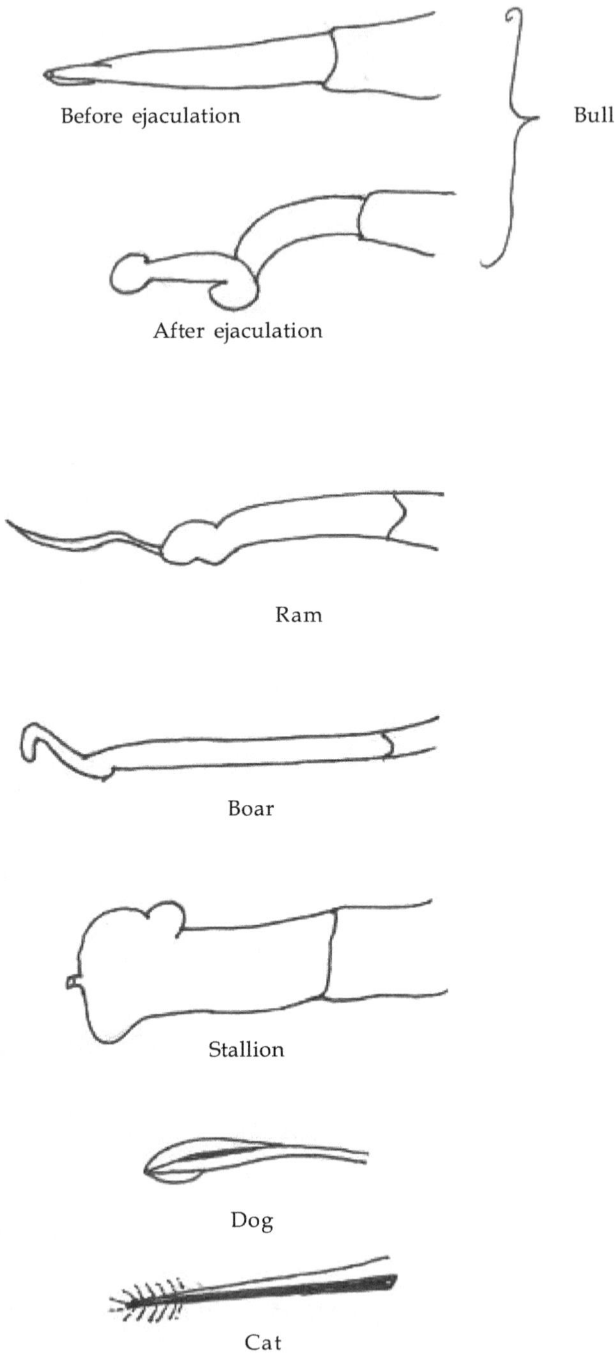

Free end of penis in Bull, Ram, Boar, Stallion, Dog and Tom Cat

Characteristics of the Prostate Secretions

- Prostatic secretions are rich in enzymes like glycolytic enzymes, proteinases, phosphatases, glycosidases, nucleases and nucleotidases.
- In dogs pH of the prostatic secretions is 6.5 and no reducing sugars are present. They, however, contains citric acid and acid phosphatase and high concentration of zinc.
- Zinc concentration in the seminal plasma is chiefly due to prostatic secretions.

1.2.2.3 Bulbo Urethral / Cowper's Glands

These are paired glands, generally ovoid shaped in bulls, stallion and rams and are large and cylindrical in boar. They are located on either side of the pelvic urethra, near the ischiatic arch. Cowper's glands are absent in dogs. In cats they are nearly as large as prostate. In stallions each bulbo urethral gland has 6-8 excretory ducts and they open into the urethra whereas only single excretory duct per gland is present in bull and boar. These glands have a diameter of 2.5 to 5 cm in stallion and comparatively slightly smaller in bulls. In bulls these glands are embedded under the bulbo spongiosus muscle (Diameter 1.5-3.0). Dribbling seen in prepuce of bulls before mounting are secretions from the prostate and bulbo spongiosus. These secretions clear harmful substances, if present in the urethra; prior to ejaculation.In Boar the glands are cylindrical, large and dense. The typical rubber like white substance filled in the cowper's glands of boar is essential for the gel formation in boar semen.

1.2.3 Copulatory Organs

1.2.3.1 Penis

It is the male copulatory organ, more or less cylindrical in form in all species. It extends forward from the ischiatic arch to the umbilical region on the abdominal wall except in cat. Prescrotally it is situated in prepuce/sheath.

Based on morphology penis is of two types :

1. *Musculocavernous type penis* : This type has large cavernous space which fills up with blood during erection e.g. horse, dog and man.
2. *Fibroelastic type penis* : This type contains lots of fibroelastic tissue with small cavernous space and a sigmoid flexure e.g. ruminants and pigs.

Mammalian penis consist of the following parts-

1. Glans- It is the terminal part of the penis lies free in the sheath.
2. Body - Composed of the large corpus caverosum penis enclosed by a thick fibrous capsule, the tunica albuginea. Ventral to this is corpus cavernosum urethrae, which is a small structure surrounding the urethra. These two structure are spongy in nature consisting of many spaces (greatly enlarged capillaries), which are continuous with the veins of penis. Distension of these spaces with blood cause penile erection.
3. Root - Root of the penis is formed by two crura that fasten the penis to either side of the ischiatic arch.

   a. Ischio cavernosus or erector penis muscleis short paired muscle that arises from the tuber ischii and sacrosciatic ligament and is inserted on the crura and body of the penis. It causes compressing and pumping action on the bulbous portion of the corpus cavernosum penis underlying the muscle leading to erection.

   b. Retractor penis muscle- This is paired smooth muscle, passes along the ventral caudal surface of the penis and attaches to the tunica albuginea of the penis. This muscle acts to draw the penis back into the sheath after erection.

Species Difference in the Penis of the Domestic Animals

1. Bulls- S-shaped curve or sigmoid flexure is present caudal and dorsal to the scrotum. During erection the S- curve is obliterated.
   Length of the penis- 36 inches (from root to tip of glans)
   Diameter- 4.5 cm (in erect state)
   Glans- 7.5-12.5 cm
2. Ram- Urethral process is present extending 4-5 cm beyond the glans penis. It has a well developed sigmoid flexure.
   Length of the penis- 30 cm
   Diameter - 1.5-2 cm
   Glans penis - 5-7.5 cm
3. Stallion - The equine penis contains large amount of the erectile tissue. Dorsal to the urethral process of the glans is the urethral sinus, or diverticulum, which is sometimes filled with smegma (called bean). In penis of the stallion sigmoid flexure is not present.

Length - 50 cm (15-20 cm lie free in prepuce)
Diameter 2.5 - 65 cm

4. Boar – The penis of the boar is similar to the bull but the sigmoid flexure is prescrotal. Glans penis is not present. Cranial portion of the penis is spirally twisted counterclock wise. Length of the penis is 45-55 cm and it protrudes 20-35 cm beyond the prepucial orifice at the time of copulation.

5. Dogs- In dogs the caudal part of the penis has two distinct corpora cavernosa separated by a median septum. The cranial free portion contains a bone , os penis, which is 5-10 cm long. Ventrally this bone is groved for urethra. Glans penis consists of two parts

   Pars longa glandis- It is the distal or cranial two third of the glans (length- 5.3 cm)

   Bulbus glandis - It is the proximal one third of the glans. It expands substantially at the time of erection and prevents withdrawl during ejaculation (Legth- 2.4 cm; Diameter - 2 cm).The length of the non erect penis in dogs is 6.5 to 24 cm.

6. Cats- In cats the penis is short and directed caudally and downwards. Os penis is generally absent in cats. In few cases, if present, it is short (3-4 mm). Urethra is located dorsally. Glans penis is also not present. A terminal cap about 1 cm long containing about 120 spines pointing towards the base of the penis is present. This is the reason for females to cry out on intromission. Bulbus glandis is absent.

1.2.3.2 Prepuce/ Sheath

It is the double invagination of skin which covers the free portion of the non erect penis and body of the penis behind the glans when penis is erect. The external opening of the prepuce is called the prepucial orifice. The fornix of sheath in the bull is the point at which the prepuce reflects upon the penis just caudal to the glans.

Species Difference in Prepuce of Domestic Animals

1. Bull- In bulls the prepuce is long and narrow about 35-40 cm long and has diameter of about 4 cm. The prepucial orifice is 5-7 cm behind the umbilicus and small with diameter of 2-4 cm. It is surrounded by turf of long prepucial hairs.Generally two pairs

of cranial and caudal prepucial muscles - protractors and retractors which acts to draw prepucial opening either forwards or backwards, are present.

2. Stallion - Stallion has prepucial cavity 15-20 cm deep and a second reflection of prepuce to form the prepuce proper of the penis. The opening between these two cavities is called the prepucial ring. Sucking noise, frequently heard, when the stallion trots is caused by engaging and disengaging of the glans penis in the prepucial ring. Smegma is the secretion of the prepucial glands and epithelial debris.

3. Ram- The prepuce in ram is similar to the bulls.

4. Boar- The prepuce of the boar has small orifice. The caudal part of the prepuce is narrow and cranial part is wide. In the dorsal wall of the wide part is an opening that leads to the prepucial diverticulum. Swine and ruminants tends to urinate inside the prepuce whereas horses, dogs and cats extends the penis beyond the sheath while urinating.

Chapter 2

Breeding Soundness Examination (BSE) of Bulls

Success of AI programme depends on fertility of bulls used for semen production. If infertile bulls are selected in semen station for production of semen doses, large population of cows inseminated with such doses is affected. Therefore, the breeding bulls or bulls to be used for frozen semen production should be selected carefully keeping in view genetic potential, disease condition, reproductive health, libido and semen production potential of the bull. The emphasis should not only be on elimination of bulls with questionable or unsatisfactory potential, but also on induction of bulls with good semen production potential into the semen station.

The following aspects should be taken into consideration while performing a breeding soundness examination of bulls:

2.1 History / Anamnesis

Past records of the fertility of the bulls should be taken into consideration while evaluating a bull for breeding soundness. The history of irregular estrus cycle, endometritis and abortions in cows bred with the bull, in past, should also be examined critically and given due consideration. Prolonged estrus interval in considerable number of cows covered with a particular bull might indicate specific infection like Campylobacteriosis and Trichomoniasis. Past illness, operations of hernia etc. should be considered before selecting bull for semen production. The collection records maintained in the semen station may also be looked into for clues of problem in the semen quality.

2.2 Physical Examination

The bull is examined in a systematic way for any problem that would hamper his ability to impregnate cows. This examination may be rather brief or more detailed if there is a reason to suspect that there is a problem with any body system. Common areas for problems are abnormalities of the feet and legs or the eyes. A bull cannot locate and mate cows unless his feet and legs are sound. Structural faults, such as sickle hocks and post legs, can cause sore feet and stresses on tendons and joints that affect the bull's mobility. Legs and joints should be free from any swelling or old injuries. Cracked hooves, corns and long hooves also slow down the breeding ability of bulls. Long hooves and corns should be dealt with, four to six weeks prior to the breeding season. This will give the bull time to recover and have sound feet before he is turned out for breeding. Eyes should be clear and free of injuries or diseases. Pink eye or cancer eye may hinder a bull's vision and reduce his breeding effectiveness. Such problems may also allow him to be dominated by other bulls and diminish his ability to cover the desired number of cows. Any other tendency toward disease or sickness should be evaluated prior to turning bulls out for the breeding season. Lumpy jaw, poor teeth, or other factors that affect a bull's ability to eat greatly reduce his breeding potential. Respiratory problems also have a negative effect on breeding ability. As a part of the physical examination a body condition score is assessed. The system used is 9-point-scale system. Bulls that are either overconditioned or underconditioned would be expected to have lower fertility.

2.2.1 Body Condition

The bull to be inducted in semen production should be properly examined for body condition. To be selected for semen production/ breeding, the bull should be healthy. It should neither be obese nor emaciated. It should have a shiny skin coat. The bull should alert and active, it should respond to external stimulus. It should be observed for any abnormal behavior and vices. The body structure, weight and conformation of the bull should be as per its typical breed characteristics.

1.2.2 Physical Examination of Organ Systems

This is very important criteria for selecting of bulls for breeding or semen production. All the body systems should be examined thoroughly and abnormalities, if any, should be considered critically.

1.2.3 Rear Leg Conformation

Rear leg conformation is vital component of a breeding soundness examination. The bulls should be examined for conformation of the rear legs and any abnormality should be considered seriously. As bulls with defective conformation of rear legs may interfere with mounting and subsequent breeding performance of the bull. The following abnormalities should be ruled out before selection of breeding bulls-

(a) Sickle hock conformation
(b) Postiness/ Post-Legged Bulls
(c) Camped behind
(d) Bow legged or narrow base or Medial rotation
(e) Cow hocked or wide base or lateral rotation or toe out stance

2.3 External Genital Examination

2.3.1 Scrotum

The skin of the scrotal area is thin, pliable and relatively hairless. The abaxial contour of the scrotum is convex. The scrotal configuration may be square, pendulous, short or twisted/ overlapping. The abnormal shape may be due to testicular hypoplasia, testicular degeneration, cryptorchidism, tumours, injury or herniation. The scrotal skin should be inspected for the presence of any inflammation, injury, insect bite or scar.

Scrotal circumference (SC): Scrotal circumference measurement is a very crucial part of the BSE. Scrotal circumference is the parameter that best predicts the output of sperm cells for bulls. There is a correlation between the scrotal circumference and the volume of semen-producing tissue of the bull. The scrotal circumference is measured with the help of a scrotal tape. It reflects the size of the testis and in turn the testicular parenchyma. The scrotal circumference, testes size and semen production are highly correlated, especially in young bulls. Scrotal Circumference has also been determined to be the one of the best predictors of bull fertility.

The scrotal circumference is measured by pushing both the testes to the bottom of the scrotum with the thumb on one side and other fingers on other side of the neck of the scrotum and then measuring the circumference at the maximum point using a tape. It is very important that the testicles be confined closely in the bottom of the scrotum for the measurement to be accurate. It is also important that the tape be closed tightly, creating a small waist on the scrotum for measurements to be consistent between evaluators. A score of 40 out of 100 is given to the scrotal circumference in breeding soundness scoring system. Normal range for the scrotal circumference in mature breeding bulls should be 32-38 cm. Since it increases with the age and weight of the bull and varies with breed it must be interpreted in light of the bull's age and breed. Inadequate scrotal circumference is a common finding in young bulls that have breeding problems. Bulls of Indian breeds and buffalo bulls have smaller testes and hence it must be kept in mind during evaluation. The scrotal diameter (SD) can be calculated by the following formula-

$$SD = SC \times 1/p$$

where SD is the scrotal diameter and SC is the scrotal circumference in cm; p is piee= 22/7)

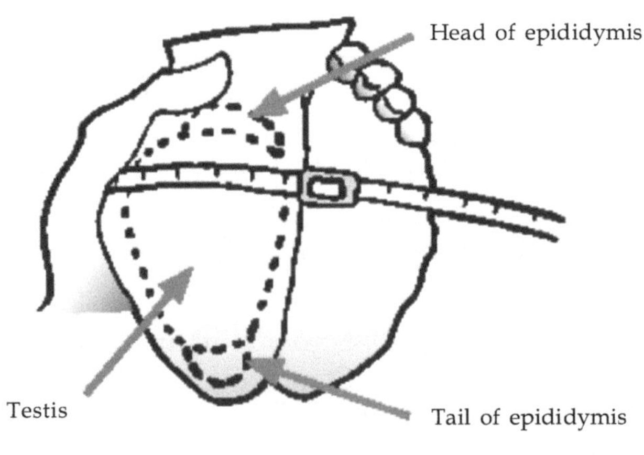

Correct method of measuring scrotal circumference in the bull. The hand restraining the testicles is placed behind the scrotum rather than to the side of it

Scrotal circumference (cm) of bulls of different age groups

Grade	Age in months			
	12-14	15-20	21-30	30 and more
Very good	>34	>36	>38	>39
Good	30-34	31-36	32-38	33-39
Poor	<30	<31	<32	<34

2.3.2 Penis and Prepuce

Anatomically penis consists of the following parts:

- *Root*: It is attached to the pelvis through crurae of penis
- *Body*: It consists of erectile tissue. The major part of the body of the penis consists of sigmoid flexure which gets straightened during erection.
- *Glans*: It is the terminal part of penis and is pointed and twisted.

The penis should be examined for the presence of any deviations in anatomy, injury or diseases. It can be best palpated at the time of semen collection or its examination typically requires the extension of the penis using an electro ejaculator.

Disorders like Phimosis, paraphimosis, large penile papilofibroma, Veneral granuloma, corkscrew shape of penis, arrest of penile development etc. should be critically examined.

The separation of the penis and prepucial sheath starts at 3-4 months of age and is normally completed at puberty. The prepuce of the bulls is long and narrow. Prepucial opening is surrounded by tuft of long prepucial hairs. Prepuce should be palpated for detection of any adhesions before selection of bulls for semen collection or breeding. Normal anatomical features characteristic of a particular breed should be taken into account during evaluation e.g. pendulous sheath is a normal anatomical feature of prepuce of indigenous bulls like Sahiwal.

2.3.3 Testis

The restrained bull should be approached from behind to palpate the testis. The size, shape and consistency of testes should be assessed. The normal length of the bull testes is 10-15 cm and the width is 5-8 cm. The smaller size of one or both the testis may indicate testicular hypoplasia and/ or degeneration. Normal testis are symmetrical and deviation from the symmetry should be carefully recorded. The normal testicular shape is oval and it should have turgid consistency. The consistency of the testis can be estimated with the help of a specially designed instrument called Tonometer. In case of normal consistency the tonometer gives a moderate deflection reading. A high tonometer deflection reading indicates testicular hardness where as a low deflection of the tonometer scale indicates softness of the testis. Flabby and soft testes may be due to testicular degeneration and hard consistency may be due to fibrosis of testes. The testicular volume can be estimated by the following formula-

$$PVT\ (cm^2) = 0.0396\ (TL)\ (SC^2)$$

Where PVT is paired testicular volume; TL is average testicular length (can be measured with Vernier calipers) and SC is scrotal circumference

The testicular volume can also be calculated according to Singh and Pangawnkar, 1989 according to the following formula-

Testicular volume = Average length of testis X total width of the two testis X Average thickness of testis

The length, width and thickness should be recorded with help of Vernier calipers or with the help of divider and centimeter scale at the points of maximum length, width or thickness, respectively.

2.3.4 Epididymis

Epididymis consists of the following parts-

(a) Caput (Head) epididymis
(b) Corpus (Body) epididymis
(c) Cauda (Tail) epididymis

The body and head of the epididymis may be palpated on the medial side of the testes by moving the other testes upside. The body enlarges at the distal pole of the testes and is termed Cauda epididymis. Tail of epididymis is normally filled and spongy.

2.4 Internal Genital Examination

To examine the internal genitalia of a bull it should be restrained properly. In furious animals, epidural anesthesia is advocated. The following structures should be palpated-

2.4.1 Pelvic Urethra

It lies on floor of the pelvic cavity and can be palpated easily as a cylindrical structure passing posteriorly downwards through the ischial arch. The anterior part of the pelvic urethra is covered partially with prostate gland.

2.4.2 Prostate Gland

The prostate gland consists of

- pars propria / Body(1.25 cm X 1.25 cm X 1.25 cm)
- pars disseminate (cryptic prostate)

It can be palpated per rectally as a ring like structure surrounding the anterior part of pelvic urethra.

2.4.3 Seminal Vesicles

They are paired lobulated glands measuring 10 cm X 5 cm X 2.5 cm in mature bulls. They are situated in the urogenital fold lateral to ampullae above the neck of the urinary bladder. In diseased conditions the lobulations are lost. There is also marked asymmetry remarkable increase in dimensions and adhesions in seminal vesiculitis. The changes in the consistency and texture of the glands are also indicative of the diseases of the seminal vesicles. Seminal vesiculitis is the most common abnormality observed in bulls. It's acute form is clinically characterized by conditions like oedema, enlargement, sensitive vesicle, pelvic peritonitis and pain on touch during per rectal examination. Conditions like generalized fibrosis, enlargement, absence of lobulation, hardness, and sharp edged vesicles may result in prolonged cases.

2.4.4 Ampullae

Ampullae are the terminal part of the vas deference. In bulls, these glands are about 10-12 cm in length and 2-2.5 cm in diameter. They lie dorsal to the neck of bladder parallel to each other and in close opposition. Ampulla is identified due to it's wide segment, softer, round and thick rubber tube like nature, which are found after manual pressure, inbetween seminal vesicles against the pelvic floor on per rectal examination. The ampullae should be examined for ampullitis.

2.4.5 Bulbo Urethral Glands

These are paired glands. Their secretions clean the urethra of deleterious effects of the urine. As they are embedded under the bulbospongiosus they are not palpable on per rectal examination.

2.5 Investigation of Libido

The libido and mating ability may be checked at the time of natural breeding or semen collection. The mating ability of a bull may be impaired by arthritis or any other causes leading to lameness. The libido can be assessed by observing the following parameters at semen collection:

2.5.1 Approach to Semen Collection Site

Observation	Score
Indifferent	0
Lethargic	1
Keen	2

2.5.2 Erection of Penis

Observation	Score
No erection	0
Within sheath	1
Outside the sheath	2

2.5.3 Reaction Time

If the reaction time is in seconds the score should be 1, if it is more than that it should be scored 0. The shorter is the reaction time, better is the suitability of the bull for breeding.

2.6 Investigation of Service Behavior

Service behavior can be examined at the time of natural service or semen collection. The service behavior can be graded on the basis of the following parameters:

2.6.1 Approach to Teaser

No interest	0
Sluggish	1
Eager	2

2.6.2 Mounting

Without protrusion	0
Incomplete protrusion	1
Complete protrusion	2

2.6.3 Location of Forelimbs

Between pin and hookbones	0
Infront of hook bones	1

2.6.4 Penile Movement

Poor	0
Weak	1
Vigorous	2
Too vigorous	3

2.6.5 Copulatory Thrust

No	0
Weak	1
Strong	2

2.6.6 Lifting of Hind Limbs from the Ground at Thrust

Nil	0
Raised	1
Complete lifting	2

A maximum score of 12 is given for each collection and average is taken after recording five collections.

2.7 Testing against Infectious Diseases

As per OIE guidelines bulls used for semen collection should be free from Tuberculosis, Johne's disease, Brucellosis, Infectious bovine rhinotrachitis, Bovine viral Diarrhoea, Campylobacteriosis and Trichomoniasis. Therefore, before inducting the bull into semen production or breeding it should be screened against these diseases and should be tested negative for all these diseases. The history of the farm from where the bull is to be procured should also be taken. Bulls should not be purchased from the farms with history of outbreaks of FMD, Brucellosis, TB etc.

2.8 Cytogenetic Investigation & Genetic Disease Diagnosis

All the bulls before being selected for breeding/ semen production should be karyotyped and any cytogenetic defect with hereditary potential should be ruled out. The bulls should also be tested for genetic diseases like BLAD (Bovine luecocyte adhesion deficiency),

Citrullinemia, Factor XI deficiency and DUMP (di uridine mono phosphate deficiency).

2.9 Evaluation of Semen Quality

2.9.1 *Semen color*: The sperms contribute the color of the semen. Opaque white or creamy white semen indicate a high concentration, a translucent while indicate medium concentration white watery semen indicate a low concentration. Blood cells in the semen indicate trauma whereas white pus is an indication of some infection in the genital tract.

2.9.2 *Semen volume*: It is highly variable between 1 to 10 ml in bulls and not usually associated with fertility.

2.9.3 *Semen pH*: The normal pH of bull semen is 6.7.

2.9.4 *Sperm motility*: The gross motility is examined by placing a small drop of neat semen without coverslip on a neat microslide and observing under low magnification. The individual motility of the semen is evaluated by examining diluted semen with coverslip under 20 X of a phase contrast microscope. This is expressed as the percent of sperms having straight forward motility. Motility has been assigned 20 points in the scoring system.

2.9.5 *Sperm concentration*: As in the case of volume of semen, the sperm concentration is also variable between 300 to 3000 million / ml. A bull with a consistent yield of sperm concentration below 600 million / ml should be investigated more carefully as it may be associated with testicular degeneration and/ or testicular hypoplasia.

2.9.6 *Sperm live and dead count*: A normal bull should have a dead count of less than 20%.

2.9.7 *Sperm abnormalities*: Society for Theriogenology recommends that bulls with more than 30% total sperm abnormalities should not be used for breeding/AI programme.

The sperm abnormalities have maximum bearing on the fertility and hence given a weightage of 40 points in the scoring system. The bacteriological examination of the semen should also be done. As per OIE standards the frozen semen doses should not contain more than 5000 colony forming units per ml. The semen samples should not contain even a single colony of pathogenic bacteria.

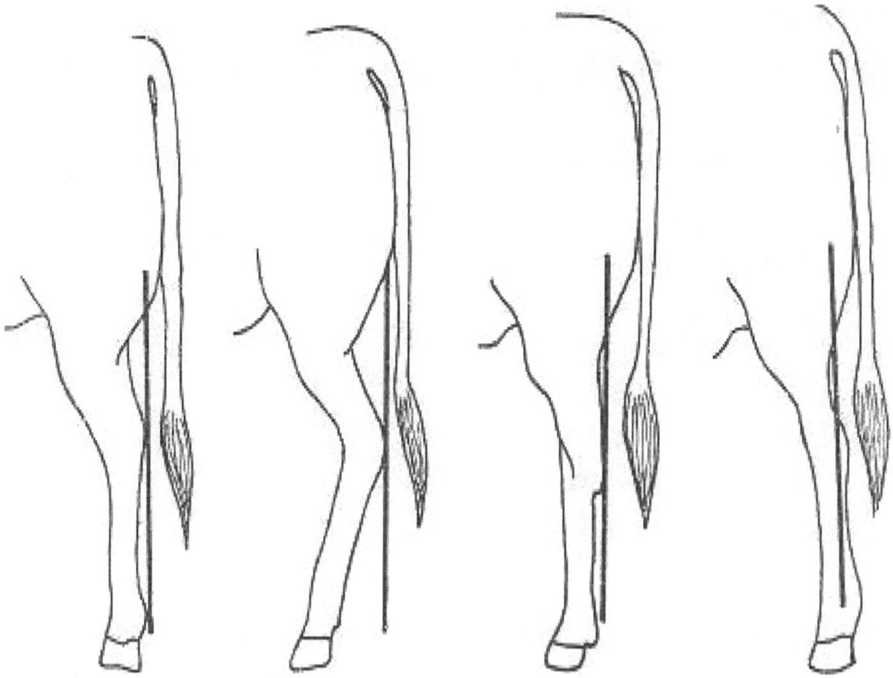

Normal | Sickle hock | Post-legged (Postiness) | Camped behind

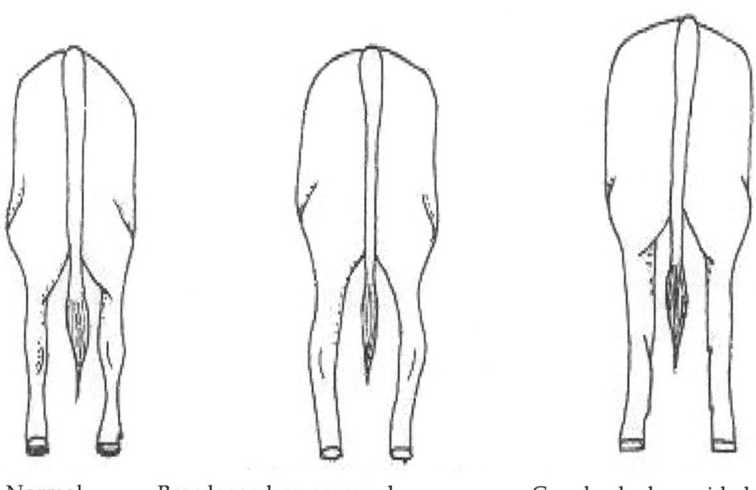

Normal | Bow-legged or narrow base or Medial rotation | Cow hocked or wide base or lateral rotation ro toed out stance

Normal and Abnormal rear leg conformation in bulls

Proforma for Breeding Soundness Examination of Bull

A. History and general investigation

1. Name of the bull rearing station................
2. Date of examination.............................
3. Name/ Number of bull..........................
4. Species & Breed.................................
5. Date of birth.....................................
6. Weight ....
7. Temperament....................................
8. Visible mucus membranes......................
9. Palpable lymph glands..........................
10. Hair Coat: Shining/ Dull/ Dehydrated/ Alopecia/ Long hairy
11. Body condition: Very Good/ Good/ Fair/ poor/ Debilitated
12. Temperature: ....
13. Conditions of legs, joints, posture and gait....................
14. Date of last vaccination:

    FMD.............HS................BQ...............
15. Record of previous deworming
16. Date of last disease screening against

    TB.........JD.........Brucellosis.........Campylobacteriosis...

    Trichomoniasis............IBR..............
17. Age at first service/ semen collection.............................
18. Date of last collection..
19. Average collection frequency during last 6 months..............
20. No. of ejaculates discarded during last 6 months.................
21. Conception rate during last 3 years...............................
22. Defects noticed, if any in progeny: Blindness, Hernia, Harelip, Cleft palate, spastic syndrome..

23. Whether reared in the farm or purchased..................
24. Muzzle and Nostrils: Moist and smooth/ rough and dry
25. Dental pad, Gums and Teeth: Normal/ Abnormal

B. Examination of different organ systems

1. Circulatory system
   a. Pulse...
   b. Percussion and Auscultation of heart.............
2. Respiratory system
   a. Rate of respiration....................................
   b. Type of respiration...................................
   c. Percussion and Auscultation of lung.................
3. Digestive System
   a. Appetite...
   b. Ruminal movements..................................
   c. Rumen flora and pH..................................
   d. Examination of faeces

   (i) Consistency
   (ii) Color
   (iii) Smell
   (iv) Parasitological examination
   (v) Presence of undigested material
4. Urinary System
   a. Palpation of the kidneys per rectum
   b. Palpation of urinary bladder per rectum
   c. Examination of urine

   (i) Volume
   (ii) Color
   (iii) Specific gravity
   (iv) Albumin
   (v) Sugar
   (vi) Blood
   (vii) pH
   (viii) Microscopic examination

C. Examination of external genitalia

1. *Prepuce and prepucial opening*: Normal/ Abnormal
   If abnormal the nature of abnormality
2. Penis: Normal/ Persistent penile frenulum/ Corkscrew penis/ Rain-bow penis /Any other abnormality
3. Scrotum

   Size: Normal/ Abnormal

   Circumference............

   Length.....................

   Condition of skin.....................

   Scrotal raphae: Present/ Absent

4. Testicles

   Present/ Absent in scrotal sac

   Right: Length..................Breadth..............Thickness..............

   Left: Length..................Breadth..............Thickness..............

   Form: Ovoid/ Accentric/ Elliptical/ any other form

   Consistency: Turgid/ Elastic/ Hard/ Fibrous/ Soft/ Fluctuating

   Movement in scrotum...

   Sensitive on Palpation...

5. Epididymis

Epididymis	Presence	Form	Consistency	Sensivity
Caput	----	-----	------	-------
Corpus	-----	-----	------	-------
Cauda	----	-----	------	-------

6. Seminal Vesicles

	Right	Left
Size	---------	------------
Consistency	---------	------------
Pain reflex	----------	-------------

7. Prostate: Normal/ Abnormal
8. Vas deference
9. Ampullae

	Right	Left
Size	----------	------------
Symmetry	---------	------------
Sensitivity	----------	------------

D. Investigation of libido

1. Approach to Semen collection site

Observation	Score
Indifferent	0
Lethargic	1
Keen	2

2. Erection of penis

Observation	Score
No	0
Within sheath	1
Outside the sheath	2

3. *Reaction time*: If the reaction time is in seconds the score should be 1, if it is more than that it should be scored 0.

E. Investigation of service behavior

1. Approach to teaser

No interest	0
Sluggish	1
Eager	2

2. Mounting

Without protrusion	0
Incomplete protrusion	1
Complete protrusion	2

3. Location of forelimbs

Between pin and hookbones	0
Infront of hook bones	1

4. Penile movement

Poor	0
Weak	1
Vigorous	2
Too vigorous	3

5. Copulatory thrust

No	0
Weak	1
Strong	2

6. Lifting of hind limbs

Nil	0
from the ground at thrust	
Raised	1
Complete lifting	2

F. Investigation of semen

1. Volume ------------ml
2. Colour ------------
3. pH ------------
4. Motility ------------ %
5. Sperm concentration ------------million/ ml
6. Live sperm ------------ %
7. Total Sperm Abnormality ------------ %
8. Bacterial Count ------------ CFU/ml
9. Pathogenic bacteria (if any) ------------

G. Investigation of infectious diseases

1. Tuberculosis : Positive/ Negative
2. Johne's disease : Positive/ Negative
3. Brucellosis : Positive/ Negative

4. Infectious bovine rhinotrachitis: Positive/ Negative
5. Bovine viral Diarrhoea: Positive/ Negative
6. Campylobacteriosis: Positive/ Negative
7. Trichomoniasis: Positive/ Negative
8. Any other ------------

H. Investigation of genetic diseases

1. Cytogenetic abnormality (if any) ------------
2. DUMP Positive/ Negative
3. BLAD Positive/ Negative
4. Citrullinemia Positive/ Negative
5. Factor XI deficiency Positive/ Negative
6. Any other ------------

Comments

The bull is Fit/ Unfit for natural or artificial breeding

Date: **Signature & Designation of the Investigator**

Place:

Chapter 3

Selection, Care, Training and Maintenance of Bulls for Breeding and Semen Production

3.1 Selection of Breeding Bulls

A sire is called half the herd and the breeding bulls to be used for semen production should be carefully selected. Selection of superior sire for breeding purpose is very important for the success any livestock breeding programme. Therefore, breeding sire must be carefully selected and maintained in scientific way so that its breeding life is prolonged and maximum number of frozen semen doses can be produced from it.

The selection criteria for future bulls should be meticulously designed keeping into account all the factors/parameters which affects the breeding programme. The selection of breeding bulls is based on three broad parameters-

3.1.1 Pedigree Examination

3.1.2 Gross Physical Examination

3.1.3 Laboratory Examination

3.1.1 Pedigree Examination

Bulls designated for breeding programme should be born of mating between proven bulls and elite dams of higher production and performance. The aim of pedigree examination is to find out if the bull is having a distinguished line of descent. An accurate pedigree selection is important for any breeding programme. The minimum pedigree standards for bulls of different breeds to be used for breeding/ semen production are given below-

Breed	Dam's best lactation yield	Fat %
Holstein Frisian	5500	3.5
Jersey	3500	5
Sahiwal	2500	4
Red Sindhi	2500	4.5
Gir	2500	4.5
Kankrej	2000	4.5
Tharparker	2000	4
Hariyana	2000	4
Rathi	2000	4
HF Cross	4500	4
Jersey Cross	3500	4.5
Murrah	3000	7
Mehsana	3000	7
Nili Ravi	3000	7
Jaffrabadi	3500	8
Surti	2000	7

3.1.2 Gross Physical Examination

Gross physical examination is extremely important criteria for breeding bull selection. This is done to find out if the bull has got any structural/ anatomical deformity which could reduce breeding/ reproductive efficiency of the bull. The physical examination should also be done to ensure that the bull confirms to its typical breed characteristics. A systematic and complete physical examination should also include checking of growth rate of bull, examination of reproductive organs and study of sexual behavior of the bulls. The size, symmetry and consistency of the testes should also be examined. The penis and prepuce should be examined to rule out the presence of any injury, growth and adhesions and other pathological conditions/

deformities. The bull should also be examined for its sexual behavior and libido. Parameters for gross physical examination of breeding bulls are described in the previous chapter on breeding soundness examination.

3.1.3 Laboratory Examination

The conventional approach of evaluation of male fertility is based on evaluation of semen samples in laboratory by few macroscopic and microscopic parameters. The functional capacity of spermatozoa cannot be predicted by these tests and there is no direct correlation of fertility of bulls with these tests. Therefore, before selecting a bull for breeding, laboratory investigation of the bulls with multiple parameters including macroscopic, microscopic and invitro fertility tests along with cytogenetic, hormonal and immunological investigation should be done. The different laboratory tests recommended for selection of breeding bulls are listed below and described in details in the section semen evaluation-

Macroscopic Examination	Appearance, Color, Volume, pH, specific gravity
Microscopic Examination	Sperm concentration, Mass motility, Progressive motility, per cent live, Sperm morphology
Functional Test	Hypo osmotic swelling test, Nuclear chromatin decondensation test, Zona free hamster ova penetration test.
Cytogenetic Investigation	Numeral and structural deviation in Karyotype
Molecular disease diagnosis	Screening for BLAD, DUMP, Citrullinemia, Factor XI deficiency and Viewer syndrome
Hormonal Parameters	Estimation of LH, FSH and Testosterone
Immunogical Parameters	Inhibin, IgA, sperm specific antibodies

3.2 Preparation of Young Bulls

Ideally bulls should be procured at 6-12 months of age. At 9 months of age a bull nose ring made of copper should be fixed. At 12 months of age the bulls should be screened for vibriosis, trichomoniasis, brucellosis, johne's disease, tuberculosis etc. They should also be screened for genetic diseases as mentioned earlier. At 18 months of age bulls will be trained for semen collection. At 24 months of age semen will be collected and evaluated based on which bulls will be selected.

3.3 Management of Breeding Bulls

3.3.1 Housing
3.3.2 Feeding
3.3.3 Disease Management
3.3.4 Restraining
3.3.5 Exercise
3.3.6 Clipping of Prepucial hairs
3.3.7 Prepucial Wash
3.3.8 Management of Stress

The management of breeding bulls is very important for harvesting the maximum semen production from it. Set of improvised management practices can significantly increase the productive life and production potential of the breeding bulls. The following aspects of bull management need utmost care if the bull has to be maintained in optimum productive stage.

3.3.1 Housing

An ideal bull pen should consist of a covered area and an open / loafing area. The dimensions of the covered and open area should be 7-8 sq meter and 12-15 sq meters, respectively for each bull. The bull pen should be length wise in rows in east west direction and breadth wise in north south direction. The covered area should have wall upto half of its height to facilitate free air movement. Height of the covered area of bull pen should be around 1.5-2 meters more than the bull's height so as to avoid radiation effect. The bull pen may be arranged in head to head pattern with a feeding passage of 3.5-4 meters so that a tractor with trolley can pass to drop concentrates and fodder. Manger for feeding and water trough should be constructed in the bull pen. Cool and potable water should be made available to the bulls round the clock. Cooling systems in the bull pen during summer months significantly improve the productivity and productive life of the breeding bulls. The floor of the covered area should be non slippery. Rough concrete flooring is best suited for bull sheds as they are impervious and easy to clean and disinfect. The concrete flooring also act as an abrasive surface and help normal wear of the hooves. Gradient of the floor should be made in such a way so that the urine and water should easily be drained. The bull pen should be cleaned, washed and disinfected daily. A foot wash should be constructed at the entrance of the bull shed so as to enable the farm personnel to dip their foot before entering the bull pen. This helps in preventing entry of infection into the bull shed. A slurry tank should be constructed near by so as to allow the accumulation of the dung and urine, which can be passed to the pastures/ fields as manure or can be better utilized as biogas.

3.3.2 Feeding

The young bulls should be fed at optimum rate so as to insure proper growth and development into a potential future sire. From 18 months to 36 months of age, the bulls should consume at the rate of 2.5 to 3.0 % of their body weight. The grain part should slowly be decreased and the roughage part should be increased as the bull grows. The young bulls can grow at the rate of 700-800 g per day and attain a body weight of 300 kg in around 12 months of age when fed optimally. During the growth phase the bulls need growth promoting ration and should be fed with adequate proteins. Thereafter, the ration should be formulated based on maintenance requirement and frequency of semen collection/breeding. The nutrient requirement of bulls is tabulated in the following :

Nutrient Requirement of growing bulls and bulls in service

Body weight (Kg)	DCP (Kg)	TDN (Kg)	Carotene (Kg)	Calcium (g)	Phosphorus (g)
Requirement of growing bulls					
100	0.28	1.9	11	13	10
150	0.35	2.6	16	13	12
200	0.40	3.0	21	13	12
300	0.45	4.0	32	13	12
400	0.48	5.0	40	12	12
Requirement of bulls in service					
400	0.38	3.6	45	9	9
500	0.45	4.5	55	11	11
600	0.53	5.4	66	13	13

Green fodder is an essential requirement of bulls. If green is not available Vitamin A should be supplemented in the ration.

3.3.3 Disease Management

Routine inspection by veterinarian to trace the bulls suffering from diseases should be done and the bulls found sick should be treated immediately. Bulls suffering from infectious/ contagious diseases should be segregated in isolation shed and treated.

Disease management is an important component of management of breeding bulls. Before induction into semen station the bulls should be screened for Brucellosis, Tuberculosis (TB), Johns' disease (JD), Infectious bovine rhinotraceatis (IBR), Trichomoniasis, Campylobacteriosis etc. Before inducting into the breeding programme the bulls should also be screened for genetic diseases like Bovine Leukocyte Adhesion deficiency syndrome (BLAD), Citrullenemia, Diuridine Monophosphate Deficiency (DUMP) Syndrome, Factor XI deficiency etc. The animals should also be Karyotyped and the bulls with abnormal karyotype should be discarded from the breeding programme as it may lead to decrease of reproductive efficiency of the progenies produced and may also act as source of spread of genetic abnormalities in the herd as some of these defect have hereditary predisposition. The newly purchased bulls should be housed in quarantine shed for a minimum of 60 days before their entry into the main herd. The quarantine shed should be located at least 5 km away from the facilities meant for the resident bulls. No equipment and manpower of the quarantine shed should be shared with the main herd. The tests for all the major diseases like TB, JD, Brucellosis and IBR should be conducted for all the newly purchased bulls in the quarantine. If any bull is found positive it should immediately be culled and the remaining bulls should be further tested twice for that disease. If any animal is found positive for that disease all the animals of the batch should culled. All the newly purchased bulls should be vaccinated against FMD, HS and BQ before inducting into the main herd.

All the bulls of the frozen semen station/ bull rearing station should be tested for infectious diseases as per the following schedule:

Brucellosis	Once in three months
Tuberculosis	Every year
Johns' Disease	Every year
Trichomoniasis	Once in six months
IBR & BVD	Once in six months

The animals found positive for these diseases should be culled.

The breeding bulls should be vaccinated for FMD, HS, BQ and Theileria to prevent the spread of infection through semen following outbreaks of the disease and also to prevent the economic losses to the semen station.

Diseases	Vaccination schedule Primary vaccination	Booster/ revaccination	Duration of immunity
FMD	4-6 months of age	3 weeks after primary and then after 4 months	6 months
HS	6 months of age	Annual	6 months
BQ	6 months of age	2 weeks after the primary vaccination and then annual	1 year
Theileria	2 months and above (don't administer any other vaccine for 8 weeks of this vaccination)	Once in lifetime	Lifetime

Control of Endoparasites: Deworming should be done once in a year. But in more prone areas twice a year deworming schedule should be followed. Normally broad spectrum anthelmintic should be used but if any animal suffers from parasitic infestation, specific treatment of the individual animal should be done based on the results of fecal examination.

Control of Ectoparasites: Infestation with ectoparasites like tick, mite, lice etc. may lead to irritation, stress and refusal to donate semen. Similarly fly/ insect bites in the scrotum can also cause inflammation, swelling and disturbances in thermoregulatory mechanism and decrease in semen volume and quality. Regular spray of ectoparasiticide on the body of the bulls as well as the floor and walls of the animals house should be done.

3.3.4 Restraining

Proper restrain is very important part of bull management as un restrained bulls may cause fatal damage to the bull handler and sometimes to themselves and other bulls of the herd. Bull nose rings should be applied to all the bulls. It is an effective tool in handling bulls as bull attendants can restrain the bulls with the help of a rope fastened to the ring. Person handling the bulls should understand their behavior and psychology. Bulls should never be beaten or threatened as it may lead to development of aggressive temperaments in them.

3.3.5 Exercise

Regular exercise should be provided to the stationed bulls preferably with the help of a bull exerciser. It helps to keep their body toned and fit. Exercise also helps to reduce reaction time, increase the semen volume and quality. It is preferable to exercise them early in the morning. The exercise should be avoided during hotter parts of the day.

3.3.6 Clipping of Prepucial Hairs

The prepucial hairs of the bulls meant for semen collection should be clipped to only 2 cm to prevent infection. Before collection the prepuce should be washed with luke warm water or normal saline and should be wiped with a clean towel or disposable paper napkin. If a towel is used use separate towel for each bull.

3.3.7 Prepucial Wash

Prepucial washing should be done with 0.01% acriflavin solution on the evening before the semen collection. This leads to significant decrease in bacterial load of the semen. Prepucial wash also prevents the contamination of semen with dust, dirt, dead cells and exudates.

3.3.8 Management of Stress

Any form of stress like thermal stress, high humidity, high ambient temperature, high radiation effect, cold and frost, shortage of drinking water etc affect the process of spermatogenesis by acting on hypothalamus-pituitary axis. It causes detoriation in the quality of semen. High ambient temperature reduces quality and quantity of semen. The ideal testicular temperature for normal spermatogenesis should be 4-5°C less than the body temperature which can be attained if the ambient temperature is 15-21°C. A constantly higher temperature can lead to abnormal spermatogenesis due to testicular degeneration. Therefore, measures to reduce the thermal stress should be the top priority in country like India where the maximum temperature in summer can go as high as 44-47°C. Particularly, bulls of exotic breeds like HF and Jersey and buffalo bulls should be housed in pens with cooling systems. The buffalo bulls should also be provided with wallowing facility.

Increased stress also leads to production of cortisol and hence reduced appetite leading to negative energy balance. Increased cortisol level may also lead to Immunosuppression and hence increased risk of infection.

Other Factors Leading to Stress and their Management

- Any disease condition, injury, pains etc. reduces the appetite leading to negative energy balance and loss of libido.
- Unsanitary conditions in bull pen, under or over feeding, lack of specific minerals, uncomfortable housing, testicular tumors, broken penis, bellanoposthitis, phymosis, abscess, arthritis, foot rot etc. may lead to stress and hence reluctance to donate semen or reduced libido.
- Too short or too tall dummy may also lead to reluctance to mount.
- Any act of inflicting pain, threat or injury etc. during collection may also lead to fear in the bull and hence distraction and refusal to mount and donate semen.
- Too high temperature in AV may burn the penis of the bull and cause fear complex in the bull and refusal to mount and donate semen.
- Some bulls may have individual preferences for a particular type of service crate or dummy and hence these factors may lead to distraction in bulls and refusal to donate semen.

Chapter 4

Semen Collection

Semen collection is the most important component of any frozen semen production programme as a high quality end product is not possible unless the raw material is of excellent quality. Therefore, all care should be taken to collect semen aseptically with the aim to harvest the best quality neat semen for processing. The person collecting semen should understand bull's behaviour and should be friendly with him. Professional bull management results in increasing sperm output and enhancing productive life of the bulls. One must know potentials and limitations of the reproductive capacity of bulls. A highly professional and dedicated team of bull handlers and semen collectors is required for this purpose. The persons involved in bull management should be familiar with bull's sexual behavior and male reproductive physiology. Preferably a veterinarian should collect semen.

Semen collection should be done early in the morning when the bulls are fresh, empty stomach in cooler atmosphere preferably between 5 to 6 am and should be finished before ambient temperature becomes high. If semen collection is done at high ambient temperature the bulls become lethargic, their libido is reduced and are exposed to stress leading to detoriation in the quality of semen. Optimally semen collection should be done twice or thrice weekly and two ejaculates a day at an interval of minimum 45 minutes (around 100 collection per year/200 ejaculates per year). Sudden changes in frequency and schedule of semen collection should be avoided.

4.1 Semen Collection Area

Most of the semen collection areas in semen station of this country lack facilities. The areas are usually small and congested and are exposed to high ambient temperature and rain. This leads to decrease in libido of bulls and quality of the frozen semen produced. The semen collection areas should be spacious so that many bulls can be collected and stimulated at a time and should have proper arrangement against extremes of temperature.

The surface of the collection yard should be soft but firm to generate adequate foot grip. Slip resistant lime stone and sand mixture (1:1) pressed in concrete groove having one feet depth is optimum flooring. Alternatively a thick rubber mat on grooved concrete floor also gives optimum results. The semen collection area should be connected to semen processing laboratory with a pass box so that the collected semen can be immediately passed to the processing laboratory.

4.2 Methods of Semen Collection

Semen collection can be done with either of the following methods:

4.2.1 Directly from vagina after natural service

4.2.2 Per rectal massage of ampulla and seminal vesicles

4.2.3 Digital manipulation

4.2.4 Electro ejaculation

4.2.5 Artificial vagina method

4.2.1 Directly from Vagina after Natural Service

In this method the semen is collected from the vagina of the female after natural service. For semen collection from the vagina a long spoon or syringe is used. The disadvantage of this method is that the semen so collected is always contaminated with mucus. It is the oldest method of semen collection and is no longer in practice.

4.2.2 Per Rectal Massage of Ampulla and Seminal Vesicles

This technique requires two person, one to do the massage and one to collect the semen. For semen collected by per rectal massage of ampulla and seminal vesicles, the bull should be properly restrained in a service crate. A well lubricated hand should be introduced into the rectum and the feces should be evacuated. The seminal vesicles and ampulla should be massaged gently pressing towards urethra for few minutes. The ampullae should be massaged and milked one by one by pressing against the floor of the pelvis. When the urethral muscle begins to pulsate the massaging action should be in synchrony with the pulsations. The semen collector must collect the cloudy fluid into a warm receptacle as it dribbles from the penis or prepuce. The extended penis may be held by the semen collector during rectal massage to facilitate collection of a clean semen sample. After massage the sigmoid flexure should be straightened to allow escape of semen, if retained there in. A urethral catheter can be used to obtain a sterile ejaculate using this technique.

Advantages

- By this method the bulls which are unable to mount due to lameness, weakness etc. can be used for semen collection.
- No expensive equipment is required.
- The technique avoids the potentially painful aspects of electroejaculation.

Disadvantages

- Skilled person is required for semen collection.
- Libido, mating ability, penile erectile function and the ability to ejaculate are not evaluated.
- Massage of ampullae may sometimes stimulate urination.
- Semen samples may be contaminated with epithelial cells, bacteria, and dirt especially when it dribbles through prepuce and hairs around the prepuce.
- Semen volume and concentration are variable.
- Some bulls may respond poorly.

4.2.3 Digital Manipulation

This method of semen collection is generally used in swine and canine wherein the males can be induced to ejaculate by applying digital pressure and massage to the penis. After the male becomes aroused, a director cone of same type, attached to a collection tube, is slipped over the penis to facilitate harvesting of semen. Some males may require to be trained for semen collection. If the male is shy having a female in estrus during semen collection can arouse it.

Semen collection from chickens and turkeys is also done by using digital manipulation technique.

4.2.4 Electro Ejaculation

This method was originally devised by Gunn (1936) for ram and was later modified for bull in 1948 by Thibault (French). Electroejaculation method was introduced by Batalli (1952) in Guinea pigs. It can be used for semen collection of almost all the mammals and wild animal (with few exceptions). While collecting semen from wild animals they need to be anaesthetized prior to the procedure. This technique can be used to obtain semen from animals that are physically incapable of mounting due to musculoskeletal disease or injury. It does not require a mount animal (dummy), and can be applied in the field using a battery-powered unit.

In this method, weak and alternating current is provided to sacral and pelvic nerves with the electrode placed in the rectum. Electro ejaculators are designed to stimulate the pelvic sympathetic and parasympathetic nerves with pulses of low voltage to induce penile erection and ejaculation. Electro ejaculators may be battery operated or plugged into an electrical outlet or may be manually operated. They may be manually operated, operate from a built in program and may be programmable; some have all three options. An electroejaculator consists of a carrying case, rectal probe, control unit, battery charger, power cord, probe cord, semen collection handle, collection cone and a collection vial.

There are also different kinds of rectal probes that vary in diameter, weight, orientation of electrodes and in the number of electrodes. These may be used according to size and breed of the bull to be collected. Probes with larger diameter produce a stronger response to stimuli of a given electrical output than smaller diameter probes. The recommended probe diameter for bulls weighing 550-900 kg is 6.5 to

7.5 cm. For larger bulls, a 9 cm diameter probe may be necessary to achieve ejaculation. For semen collection, bulls are restrained in a chute with good footing. Sometimes, it may be necessary to prevent a bull from lying down during electroejaculation by applying a restraining belt under the bull's chest. Provide a gradually increasing alternated current (from 0-5 Volts) and later reduce it from 5 to 0 volts in 5-10 seconds. The current frequency in this method should range from 15 to 90 cycles per second. Increase subsequent stimulations slowly. Normally erection and ejaculation should occur at 10-15 volts when 0.5-1.0 ampere current is flowing. It takes normally 3-5 minutes for semen collection.

It appears to be extremely painful procedure but normally does not cause much distress on bulls provided they are securely restrained. It is conducted under anesthesia in many species.

Collection Technique

Move the bull to the chute. Set up the electroejaculator beside the bull leaving it turned on and ready to be used. Using a palpation sleeve, the rectum is emptied of feces and a longitudinal massage is applied over the ampullae and urethralis muscle for 1-2 minutes. A lubricated probe is introduced into the rectum with the electrodes facing ventrally. Make sure the electroejaculator is turned on before performing this step. If it is turned on after inserting the rectal probe, the bull may receive a strong electrical pulse that will increase the level of stress in the animal.

Electrical stimulation is begun slowly until the bull shows a minimal response. Consecutive stimuli are then given, with a small increase in intensity each time. Stimuli should last 1-2 seconds and then be discontinued for 0.5-1 second before the next one starts. After several stimulations, clear pre-seminal fluid begins to flow from the protruded penis. This clear pre-seminal fraction should not be collected. As soon as the cloudy sperm rich fraction begins to flow from the penis, a collection cone with the test tube is placed over the penis and the sample is collected.

After collecting a suitable sample, the stimulation is stopped and the rectal probe is removed. The semen sample is then taken to the laboratory for evaluation and processing. While performing the procedure it is important to obtain penile protrusion for examination of the penis and prepuce. The majority of the bulls emit semen without excessive stimulation. However, if a bull has not ejaculated after reaching the highest level of stimulation, 3-4 stimulations in the maximum level can be done, followed by a rest period of 1-2 minutes with the probe still inside the rectum. Often on a second attempt to electroejaculate bulls, the penis will not protrude; therefore, while the bull rests, the penis should be held manually with a gauze sponge to prevent retraction into the preputial cavity. All four fingers with the gauze should wrap the glans penis. Do not attempt to catch the penis over the preputial region, as it will roll off into the prepuce. Semen emission commonly occurs during the rest period thus the collector should be attentive to catch the emission. After resting the bull, stimuli beginning at two voltage increments below the maximum are begun in a second attempt to obtain an ejaculate. This is often successful.

Bulls should be sexually rested for 1-2 days before electro-ejaculation to allow sperm to accumulate in the ampullae. With good equipment and proper technique, only about 2% of normal fertile bulls fail to emit semen by electroejaculation.

Disadvantages

Electroejaculation without anesthesia is painful. Vocalization during electroejaculation, an elevation of circulating adrenal progesterone and cortisol after electroejaculation are evidence of pain. Therefore, the procedure must only be done by personnel with proper training and always in the gentlest possible way.

4.2.5 Artificial Vagina (AV) Method

It is the simplest method of semen collection and it overcomes various disadvantages of all the above mentioned methods as AV is simple and easy to assemble. Complete ejaculate can be obtained in cleanliest way by this method. Considering the edge of this method over other methods it is more commonly used. Artificial vagina is used to collect semen from cattle, buffalo, horses, sheep, goats and even cats.

Types of Artificial Vagina

Russian Model

It consists of rigid rubber cylinder of 60 cm length and 5.5 cm inner diameter. Thin walled rubber tubes are present inside the cylinder, the ends of this tube were turned back over the outer cylinder to form a water tight jacket. At the time of semen collection the warm water is filled in through the screw plug hole on the cylinder. A graduated semen receptacle of glass of slightly smaller diameter than the cylinder is fitted at one end of the AV.

English Model

It is a modification of the Russian model where the semen collection receptacle has larger volume to accommodate the air pressure built when the bull ejaculates.

Cornell Model

The semen receptacle in all other models was exposed to light and cold weather and hence the semen may be exposed to cold shock. So this model has semen receptacle inside the rubber cylinder. Here the funnel was long and ran inside the whole length of the inner liner and was attached to semen receptacle. The outer rubber cylinder in this model was as long as 71.2 cm.

Danish Model

This consists of an outer rubber cylinder of 40.7 cm length and of 5.7 cm inner diameter and a rubber cone attached to a semen collection tube. The length of the cylinder depends on the size of the penis of the bulls as it is desirable that the bull should ejaculate near the cone to avoid the thermal shock to the spermatozoa while in contact with inner liner at temperature of 42-45 oC and wastage of semen while passing through the liner to the cone and subsequently the collection tube.

Preparation of Artificial Vagina

The AV consists of a rigid hard rubber cylinder, inner liner made up of soft rubber, a latex cone and a graduated glass tube. The inner diameter of the cylinder is 6 cm. The cylinder of AV for bulls is available in two different lengths of 30 and 40 cm. The AV should be selected for individual bulls based on the length of their penis. Goat AV (20 cm length) is ideal for collection of semen from buffalo bulls as the length of their penis is comparatively smaller than cattle and using bull AV in case of buffalo bulls may cause ejaculation in the liner and hence thermal shock to the spermatozoa due to high temperature in the liner and loss of part of ejaculate which sticks to the liner while traversing through it to the collection tube.

All the parts of the AV should be clean, sterile and dry before assembling or assembled AV can also be sterilized. The rubber lining of either latex rubber or PVC is mounted on the cylinder and is fixed with help of a rubber straps. Latex rubber cone is mounted on the cylinder to the end where water and air inlet are closer. The glass semen collection tube is fixed to the conical end of rubber cone and an insulation bag is mounted on glass tube and rubber cone and fixed on the rubber cylinder with strap. The collection tube should also be warmed to body temperature to avoid cold shock to the collected spermatozoa, especially in the winter months/ cold climates. The assembled AV is filled with water at proper temperature and pressure. Normally the temperature of the AV should be 42 to 44°C. The temperature required in AV varies with between bulls. Once the AV is filled with hot water, the pressure of the AV should be adjusted with air with the help of a cycle pump or rubber bulb for filling air. Pressure in the AV after filling air should be 45-55 mmHg. After filling air the open end of the AV should have appearance of a star. After filling the optimum amount of air, the air valve of the AV is closed. The anterior part (3-5 cm) of the AV should be lubricated with sterile jelly which is non toxic to the spermatozoa.

Preparation of Bulls for Semen Collection

Prepucial hairs should be clipped. The long hairs in the prepuce should be clipped as it aids in maintaining cleanliness. However, care should be taken not to clip them too short as too short hairs may induce irritation. The bull should be washed thoroughly. The washing should be done long enough before the semen collection so that the bull is dry at the time of semen collection as wet underline is conducive for bacterial contamination. The prepucial area is wiped with a clean towel/napkin soaked in 70% ethanol immediately before semen collection.

The bull should be provided with proper stimulation before collecting semen. The stimulation process starts primarily with visual and auditory stimulation, wherein the bull is exposed to the collection environment where the other bulls are being stimulated and collected. The presence of other animals and auditory stimuli causes sexual stimulation leading to erection and desire to mount. The collection arena where many bulls are stimulated and collected creates a sense of competition among the bulls adding to increased level of stimulation.

The stimulation time (maximum 8-10 minutes) should be fixed for individual bulls based on their reproductive behavior. The sexual preparation of the bulls should be done by giving three false mounts, each followed by two minutes of restraint. This results in increased quality and quantity of semen. If the bull is not prepared properly the volume of seminal plasma could be less. Seminal plasma is helpful in nourishment, activity and maintenance of fertility. It also helps in proper flushing of the reproductive tract with seminal plasma and hence increasing the sperm output.

Semen collection can be done by AV method by allowing the bull to mount on other animal termed 'dummy' after prior stimulation.

Before Starting Semen Collection from a Bull Please Ensure that

- The AV used for semen collection is sterile and proper temperature (42-44 °C) is maintained in the AV. The choice of temperature in fact depends on individual bulls. The individual preferences should always be kept in mind.
- Prepucial wash with mild antiseptic solution, preferably 0.01 % acriflavin or normal saline solution, is done 12 hours before semen collection.
- A clean bull apron should be tied to the bull prior to semen collection. It reduces wastage of ejaculates due to rubbing of penis against dummy's hind quarter. The ejaculates collected from bulls with bull apron are cleaner. Use of bull apron also prevents cross infection from bull to bull. It also permits the bull to have free penile movements for seeking vagina in natural process.
- Bull is exercised regularly as per prescribed schedule prevailing in the semen station.
- Older bulls may require rough liner and higher temperature for ejaculation. Young bulls normally ejaculate in smooth liners and normal temperature range of 42-44 oC. Rough liner and high temperature of AV should not be normally used in young bulls.

Age for Starting Semen Collection in Different Animal Species

Exotic cattle bull	12 months
Indigenous cattle bull	24 months
Buffalo bulls	24-30 months
Stallion	24 months

Different parts of Artificial Vagina

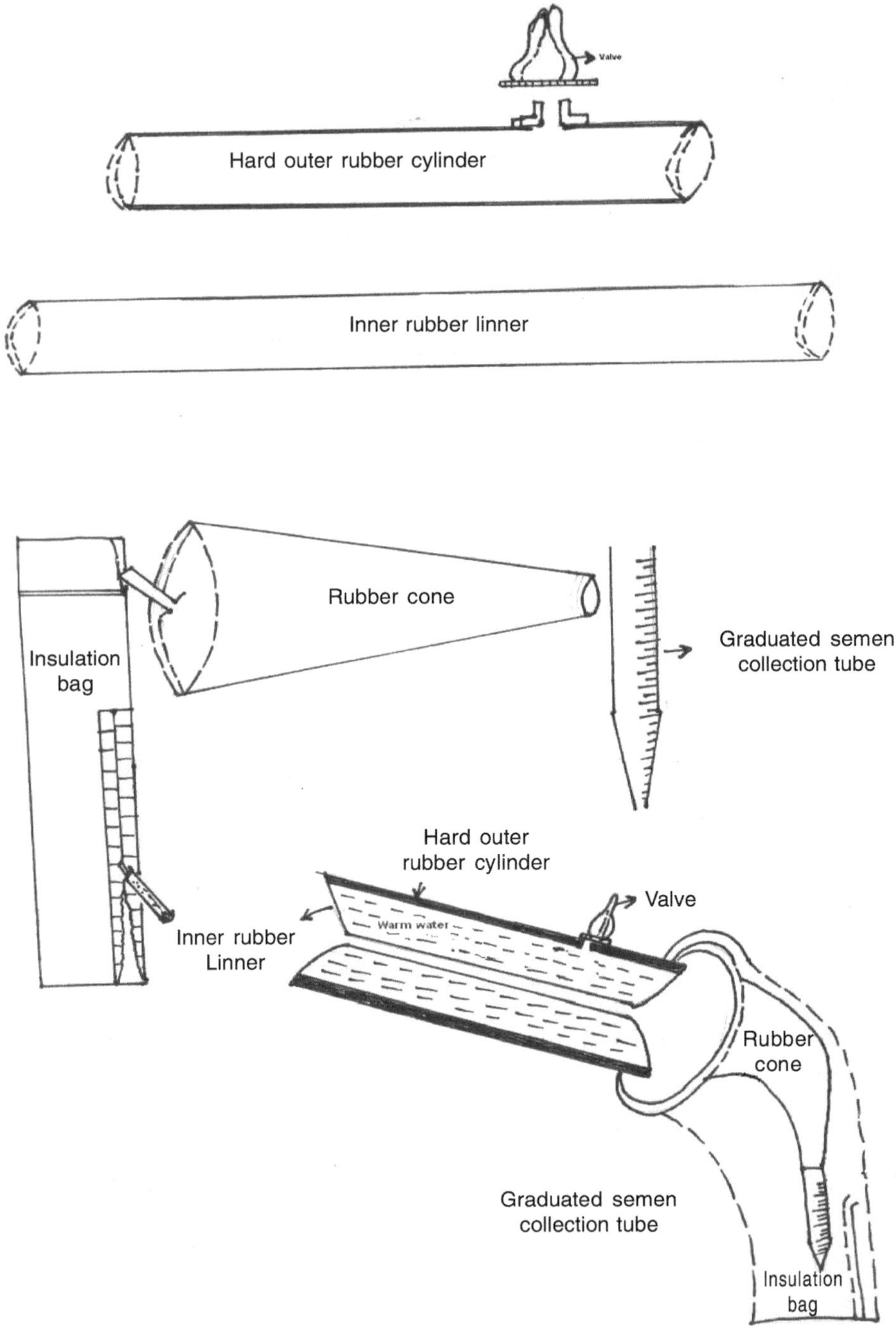

Assembled Artificial vagina

Outer hard rubber cylinder

Air blower

Inner latex liner

Insulation bag

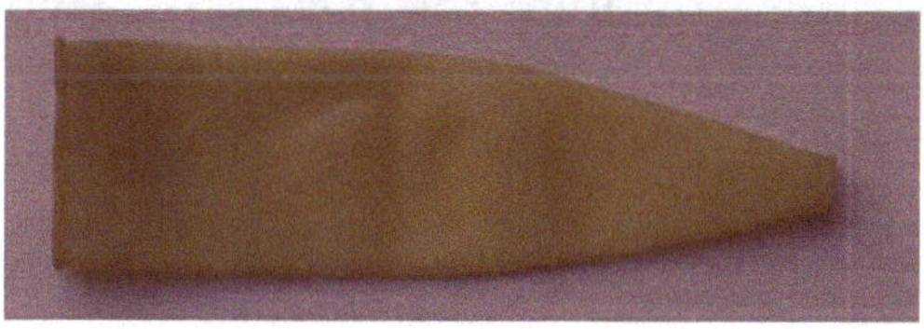

Latex Cone

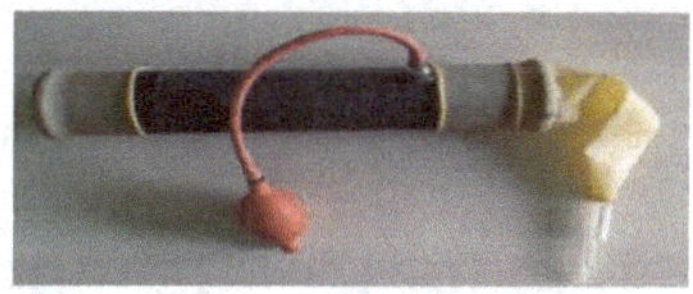

Filling air in the Artificial Vagina

Components of Artificial Vagina

Ram	7-8 months
Buck	7-8 months
Boar	7-8 months
Camels	36-40 months

Semen Collection Procedure with Artificial Vagina

Bring the bull to semen collection area. The dummy should be properly washed and tied in collection crate and restrained properly. The bulls may sometimes have their individual preference for a particular dummy. Due consideration should be given to bulls individual preferences. The bull should be given prior stimulations by giving 2-3 false mounts. Each false mount should be followed by two minutes restraint. This increases semen volume and quality. The time commencing from the release of the bull to the first mount is called the Reaction time. The lesser is the reaction time, the better is the libido of the bull. The penis of the bull should be deviated to avoid the contact with teasers posterior part if bull apron is not used. After false mounts the bull should again be restrained for few minutes before allowing them to serve the AV. This increases ejaculate volume and total number of sperm cells in the ejaculate. The semen collector should stand on the right side of the bull with AV in his right hand. He should not be afraid of the bull as the attention of the bull at the time of semen collection is on dummy and not on the semen collector. The semen collector should stand near the bull at the time of collection because if he comes running from a distance, bull may be distracted and dismount by fear of being charged or attacked. When the bull mounts on the dummy after proper erection of the penis, the penis is deflected by grasping the sheath into the AV. The glans is allowed to touch the liner of the AV. The AV should not be thrust onto the penis rather the bull should be allowed to exert thrust. The AV should be withdrawn as the bull dismounts. The semen collection tube should be disconnected from the AV and passed to laboratory through the pass box. Two ejaculates can be collected from a bovine or bubaline bull on the same day, ideally 45-60 minutes apart, for frozen semen production.

4.3 Tips for Maintaining Sex Drive of Bulls under Semen Collection

- Routine semen collection procedure in the same settings using same teaser may induce reduction in sex drive of some bulls. This can be managed using the following tips.
- Change the teaser animals for bulls which become sluggish. Sometime changing position of the teaser in restrain crate by few feet is sufficient to renew the bull's interest.
- Changes in surroundings of semen collection may be helpful in stimulating interest of the bull in semen donation. The changes may include changing semen collection area from indoor to outdoor, changes in collection personnel, their outfits etc.
- The semen collector and other personnel present in the semen collection area should avoid sudden movement and any violent activity during semen collection.
- Never mishandle or beat the bulls as mishandled, excited or distracted bulls often tend to lose interest in semen donation. Operators should wear dark and dull colored clothing as they are less distracting to the bull when movements are made. Noise or other form of disturbances should be avoided in the semen collection area.
- Any injury, pain etc. during semen collection may frighten the bull and may cause loss of interest in semen donation.

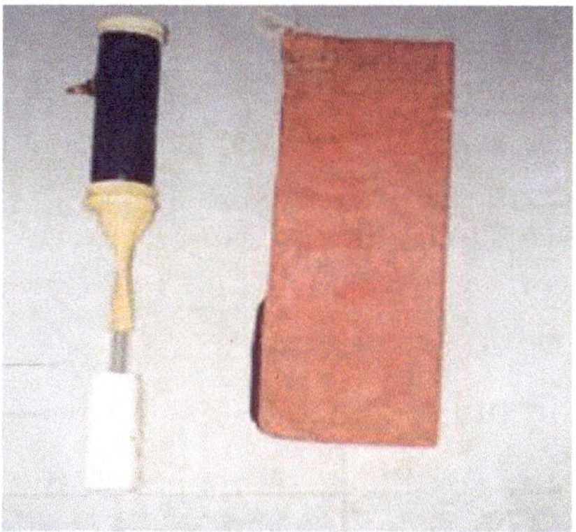

Prepared AV

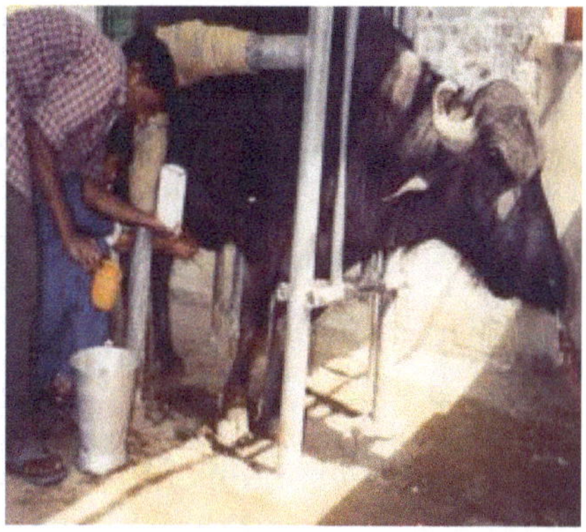

Prepucial wash

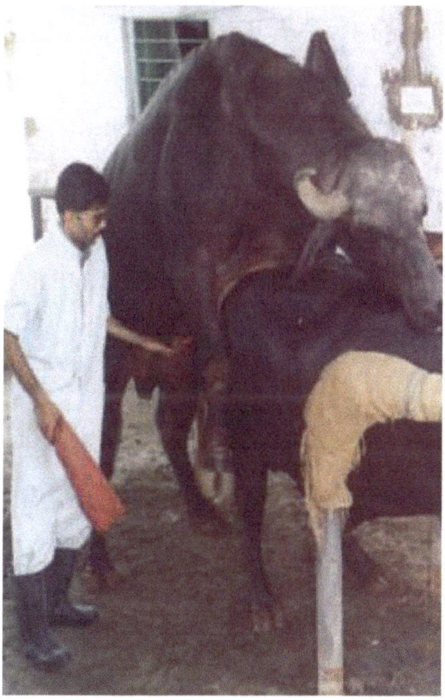

Deviating penis during false mount

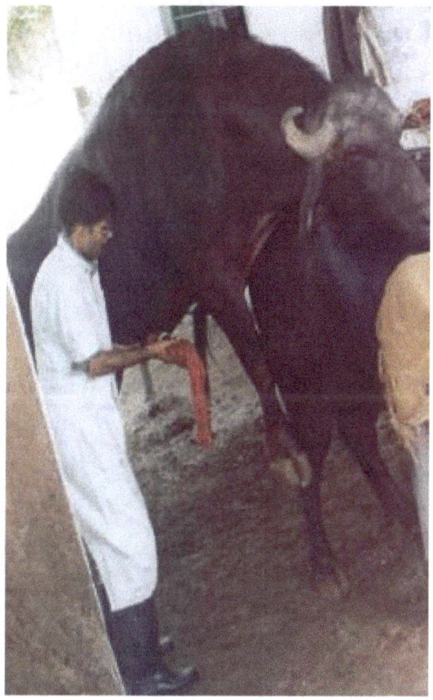

Semen collection

- The temperature of the AV should be correct. Excessive temperature causing burn may be remembered by bulls and lead to refusing to serve the AV and may cause a willing bull to become slow or non working. Low temperature may also cause failure to ejaculate. Little variation in temperature as per individual choice of the bull is, however, permissible.

4.4 Semen Collection Frequency

Semen collection frequency of the bulls depends upon the bulls ability to ejaculate normally and regularly and their semen production rate. Spermatozoa take 41 days to reach caput epididymis and 11 days to pass through the epididymis. Entire spermatogenesis comprises 4-5 cycles of 13 days each. Testicular reserves are equivalent to 3 days sperm production.

Daily sperm output is 4-81 million in bulls.

Frequent semen collections reduce the sex libido, semen volume, initial motility and density greatly. Generally semen is collected at 48-96 hours interval. It is adviced to collect semen three times a week. This collection frequency has been reported to have no adverse effect on semen quality. Bulls, however, occasionally may refuse to ejaculate or take greater reaction time.

Daily semen collection can, however, be done from outstanding bulls during active breeding season for upto 15 weeks with no difference in semen quality. It is, however, advisable to follow thrice a week collection schedule for prolonged reproductive life of the bull.

Chapter 5

Collection of Samples for Diagnosis of Reproductive Disorders in Bulls

For the diagnosis of clinical reproductive problems it is essential to collect blood, penile swab, Prepucial swab, prepucial washing, semen, smegma etc from breeding bulls. These samples are dispatched to the nearest laboratory for clinical diagnosis.

5.1 Collection of Blood Serum

The bull is properly restrained and jugular vein area is cleaned and sterilized with 70% alcohol. An sterile hypodermic needle is used to collect 5-10 ml blood aseptically in a sterile test tube. This test tube is then kept in a slanting position. Serum is separated after the formation of blood clot and dispatched to the laboratory.

5.2 Collection of Seminal Plasma

Semen sample is collected as per the standard procedure. Carbolic acid or 1 % sodium acid is added to preserve seminal plasma at the rate of 2 drops in 1 ml ejaculate in the centrifuge tube. It is then centrifuged at 3000 rpm for 20 minutes. The supernatant is collected and stored in deep freeze till further analysis.

5.3 Prepucial Wash for Mycotic/ Bacteriological Examination

Bull is retrained properly and sterilized rubber tubing is inserted into the prepucial cavity. The prepucial opening is closed tightly or ligatured. Around 200-300 ml of sterilized saline is infused into the prepucial cavity and gentle massage is done on it for 3-5 minutes and then the fluid is collected into a container.

5.4 Semen for Mycotic Examination

Semen is collected aseptically in an sterile tube as described under the section semen collection and it is stored at 4 ^{0}C and dispatched to the laboratory.

5.5 Prepucial Swabing for Parasitological Examination

A swab sponge is attached to a wire loop and wrapped in an aluminum foil and autoclaved. The penis of bull is protruded out after properly restraining the bull and putting pudendal nerve block. The swab is then saturated with saline solution and then injected into prepuce with care. The prepuce, penis and swab are pressed gently by movement with hand. Then this swab is subjected to parasitological examination.

5.6 Prepucial Wash for Parasitological Examination

The procedure is same as for mycotic/ bacteriological examination.

5.7 Smegma for Parasitological Examination

The bull is restrained properly and a long pipette attached with a rubber bulb is inserted into the prepucial cavity. Smegma is collected from the glans penis by washing with normal saline and then dispatched to the laboratory.

Section B

Semen Processing & Cryopreservation

Chapter 6

Semen Dilution and Preservation at Different Temperatures

Semen dilution/ extension is done to meet the following objectives

- To preserve the fertilizing potential of semen for a long period of time
- To increase the number of services per ejaculate. Ideally approximately 500 cows/ buffaloes can be inseminated from diluted/ extended semen of a single ejaculate.
- Extenders provide the spermatozoa with a medium in which they are able to remain viable and fertile for a prolonged time period.

In order to be an ideal extender, it should be able to provide conducive environment to prolong sperm survival and modify the sperm structure to prepare them for freezing through the following properties:

- It should be isotonic with blood and should maintain this during preservation.

- It should have capacity of maintenance of pH (6.6-6.8) with high buffering capacity in that range.
- It should ensure supply of metabolites for nutrition and survival of sperm cells.
- It should contain lipoprotein and lecithin, minerals in adequate quantity and should contain substances to be used for aerobic metabolism by the spermatozoa.
- It should provide protection against cryoinjuries.
- It should prevent the formation of ice crystals.
- It should not contain any material which may exert any adverse effect on spermatozoa, female genitalia, fertilization process or growth and development of the zygote.
- It should contain antibiotic in sufficient quantity to check the microbial growth.

6.1 Commonly Used Extenders for the Preservation of Semen

6.1.1 Preservation of Semen at Ambient Temperature (18-30 °C)

The following extenders are used for preservation of semen at ambient temperature:

A. Coconut Milk Extender
B. Milovanov's Extender
C. Illini Variable Temperature (IVT) Extender
D. Cornell University Extender (CUE)

A. Coconut Milk Extender

Buffer, Antibiotics and Catalase (Solution I)

Sodium Citrate dihydrate	2.2 g
Penicillin G. sodium	60.0 mg
Dihydrostreptomycin sulphate	135.0 mg
Sulphanilamide	300.0 mg
Polymyxin B sulfate	10.0 mg
Aqueous solution of catalase	15000 units
Mycostatin	1000 units
Triple glass distilled water	60-70 ml

Coconut water (Solution II): Collect coconut water from green coconut in a sterile glass beaker. Boil it for 10 minutes. Filter and cool to room temperature. Add 15 ml of this to buffer prepared as above.

Extender: Add 7 ml egg yolk to the mixture of Solution I and II. Make the volume up to 100 ml with sterile triple glass distilled water. Adjust the pH to 7.4 with 10 % NaOH solution.

B. Milovanov's Extender

Solution I	
Potassium dihydrogen phosphate	0.720 g
Triple glass disilled water upto	100 ml
Solution II	
Sodium citrate dihydrate	20.276 g
Glucose	5.7 g
Sodium bicarbonate	1.26 g
Sulphanilamide	3.0 g
Triple glass distilled water upto	900 ml

Mix solution I and solution II and add 11 % egg yolk. Stir gently with a sterile glass rod. One lac units of Penicillin G sodium and 100 mg of dihydrostreptomycin sulphate are added to 100 ml of extender before use.

C. Illini Variable temperature (IVT) Extender

Sodium citrate dihydrate	2.0 g
Sodium bicarbonate	0.21 g
Potassium chloride	0.04 g
Glucose	0.3 g
Sulphanilamide	0.3 g
Triple glass distilled water upto	100 ml

Carbon dioxide gas is bubbled through the above mentioned solution until the pH reaches 6.3. Dihydrostreptomycin sulphate @ 1000 mg and Penicillin G. sodium @ 1000 IU/ ml and 10 % egg yolk are added fresh at the time of use. The diluted semen is stored in 1 ml ampoule at room temperature. Ampoules should be flushed with CO2 immediately before filling.

6.1.2 Extenders used for Preservation of Semen at Refrigeration Temperature (4 °C)

For preservation at refrigeration temperature semen is diluted with suitable extender at 35 °C. Such semen samples are placed in a wide mouth glass tube with a stopper and the tube is placed in a beaker containing water at room temperature. The beaker along with semen tube is stored in refrigerator maintained at 4 °C. Such preserved semen samples can be stored upto 3-4 days. The following extenders can be used for preservation of semen at refrigeration temperature (4 °C).

A. Egg yolk Citrate (EYC) extender
B. D2 Dilutor
C. Egg yolk Glucose Bicarbonate extender (EYGB)

A. Egg Yolk Citrate (EYC) Extender

EYC is the most frequently used extender for the dilution of semen at refrigeration temperature (4 °C). This extender was first developed by Salisbury in 1941 at the Cornell University. In this extender lecithin or similar phospholipids present in the egg yolk has a protective action on the sperm cells and also provides nutrition to the sperm cells. Sodium citrate acts as a buffer and a chelating agent which binds up calcium and other heavy metals and disperses the fat globules in the egg yolk so that there is homogenous mixing of the spermatozoa in the extender and individual spermatozoa can be observed clearly under a microscope.

Method of Preparation

Buffer	Sodium citrate	2.9 g
	Triple glass distilled water upto	100 ml

Adjust the pH to 6.9. Add egg yolk to the Sodium citrate buffer in the ratio 1:4. Add Penicillin G sodium @ 50-100 IU/ml and Streptomycin sulphate @ 50-100 µg/ ml to the extender at the time of use. Mix well with a magnetic stirrer.

B. D 2 Extender

Sodium bicarbonate (1.3 % solution)	10 parts
Glucose (5 % solution)	40 parts
Fructose (5 % solution)	25 parts
Egg yolk	25 parts

One lac units of Penicillin G Sodium, 100 mg of Dihydrostreptomycin sulphate and 300 mg of Sulphanilamide are added to 100 ml D 2 extender before use. This extender is good for the preservation of buffalo semen.

C. Egg yolk Glucose Bicarbonate Extender

Solution I

Sodium bicarbonate	0.325 g
Dihydrostreptomycin sulphate	0.18 g
Penicillin G. Sodium	0.075 g
Triple glass distilled water upto	25 ml

Solution II

Glucose	5 g
Sulphanilamide	0.375 g
Triple glass distilled water upto	100 ml

Add 25-33 % egg yolk. Mix with magnetic stirrer. This extender is good for preservation of buffalo semen.

6.1.3 Extenders used for Cryopreservation of Semen (-196 ^{0}C)

As a principle all extenders which can be used for dilution and preservation of cattle and buffalo semen at refrigeration temperature can be used for cryopreservation of semen of corresponding species of animal on addition of a suitable cryoprotectant. The following extenders are frequently used for cryopreservation of cattle and buffalo semen.

A. Commercial extenders

B. Tris citric acid egg yolk extender

C. LFYG (Lactose, Fructose, Egg yolk, Glycerol) extender

D. Egg yolk citrate extender (EYC)

E. Skim milk or whole milk extender

A. Commercial Extenders

The commercial extenders are available in ready to use state after dissolving in distilled water. Various preparations like SPV-161, Spermasol plus, 20 % egg yolk, Seminan, Laciphos-123 with 10 % egg yolk are available in the market.

B. Tris Citric Acid Egg Yolk Extender

This is the widely used extender for cryopreservation of cattle and buffalo semen. The method of preparation of this extender is described in details in the next section.

C. LFYG (Lactose, Fructose, Egg yolk, Glycerol) Extender

Lactose 11 %	56.25 ml
Fructose 6 %	18.75 ml
Egg Yolk	20.00 ml
Glycerol	5.00 ml
Penicillin G. Sodium	1000.00 IU/ ml
Streptomycin	1000 μg/ ml

D. Egg Yolk Citrate Extender

Buffer	
Sodium citrate	2.9 g
Glycerol	7.0 ml
Triple glass distilled water upto	100 ml

Adjust the pH to 6.9. Add egg yolk to the Sodium citrate buffer in the ratio 1:4. Add Penicillin G Sodium @ 50-100 IU/ml and Streptomycin sulphate @ 50-100 μg/ ml to the extender at the time of use. Mix well with a magnetic stirrer.

E. Skim Milk or Whole Milk Extender

Skim milk or whole milk heated at 92 °C for 10 minutes	74.0 ml
Glycerol	6.00

Add 20.0 ml egg yolk, Penicillin G Sodium @ 50-100 IU/ml and Streptomycin sulphate @ 50-100 μg/ ml to this extender at the time of use. Mix well with a magnetic stirrer.

Calculation of Volume of extender for Semen dilution

Let the ejaculate volume	=	2.0 ml
Sperm concentration	=	1000×10^6 / ml
Progressive sperm motility	=	70 %

Contd...

Contd...

Each ml of neat semen contains $\frac{1000' \, 10^{6}' \, 70}{100}$ progressively motile sperm

= 700 X 10^6 progressively motile sperm.

Therefore, 2 ml semen contains 2 X 700 X 10^6 progressively motile sperm

= 1400 X 10^6 progressively motile sperm

Dilution Rate for Liquid Semen

Each insemination dose should contain 10 X 10^6 progressively motile sperm

Therefore, Number of doses to be produced = $\frac{1400' \, 10^{6}}{10' \, 10^{6}}$

= 140 doses of 1 ml each

Volume of neat semen = 2 ml

Volume of extender to be added = 140- 2 = 138 ml

For Frozen Semen

Each insemination dose should contain 20 -30 X 10^6 progressively motile sperm (let's consider 20 X 10^6 progressively motile sperm/ insemination dose)

Number of doses produced if semen is diluted to 20 X 10^6 progressively motile sperm / ml

= $\frac{1400' \, 10^{6}}{20' \, 10^{6}}$

= 70 ml

For 0.5 ml insemination dose semen containing 20 X 10^6 progressively motile sperm per insemination dose semen should be diluted to 70 x 0.5 ml = 35 ml

Volume of the neat semen = 2 ml

Volume of the extender required = 35-2 = 33 ml

For 0.25 ml insemination dose semen containing 20 X 10^6 progressively motile sperm per insemination dose semen should be diluted to 70 x 0.25 ml = 17.5 ml

Volume of the neat semen = 2 ml

Volume of the extender required = 17.5-2 = 15.5 ml

Chapter 7

Preparation of Tris Citric Acid Egg Yolk Glycerol Extender and Processing of Semen for Cryopreservation

The Tris citric acid egg yolk glycerol extender is the most frequently used extender in the cryopreservation of cattle and buffalo semen. Preparation of the extender plays a very important role in determining the quality of the processed semen. Therefore, care must be taken to ensure that the extender is prepared aseptically and it contains all the ingredients in appropriate concentration. Extender should always be prepared under a laminar air flow. It is also important to check that the pH and osmolarity of the semen extender for a particular species corresponds to the prescribed limits for that particular species. The Tris citric acid egg yolk glycerol contains the following ingredients:

Protective agent	Egg yolk, Glycerol
Sugar	Fructose
Antibiotics	Penicillin and Streptomycin
Buffer	Tris (tris - hydroxymethyl amino methane)
	Citric acid monohydrate

Cryoprotective Action of Glycerol

Glycerol exerts cryoprotective effect by the following ways:

- It binds with water and decreases the freezing point of the solution and acts through 'salt buffering mechanism.
- It inhibits lactic acid production and prolongs sperm survival.
- It enters the sperm cell and binds some of the intracellular water and protects sperm from mechanical damage by intracellular as well as extracellular water crystal formation and changes in concentration of electrolytes.

7.1 Method of Preparation of Tris Citric Acid Egg Yolk Glycerol Extender

Solution A

Tris	3.028 g
Citric acid monohydrate	1.675 g
Glycerol	7.00 ml
Triple glass distilled water upto	100 ml

This solution is sterilized by autoclaving at 5 p.s.i. pressure for 20 minutes.

Solution B

Fructose (anhydrous)	1.25 g
Triple glass distilled water	20 ml

This solution is sterilized by Tyndallization.

The 80 ml of solution A and 20 ml of solution B are mixed. Add Benzyl Penicillin Sodium (1 Lakh IU) and Streptomycin sulphate (100 mg) to prepare Tris buffer. This buffer can be stored at 4 °C for 3-4 days.

Tris Extender

Tris extender is prepared by adding 20 % egg yolk to Tris buffer. This should be prepared fresh at the time of use and thoroughly mixed using magnetic stirrer. It should be maintained at 34 °C at the time of semen processing.

Preparation of Egg Yolk

Egg yolk is prepared from fresh hen's egg procured from a known source / farm (free from infection). Procurement of egg from market should be avoided as there are chances of microbial contamination. Eggs should not be stored for more than 2-3 days at 4 °C. Eggs should never be cleaned by washing with water as there is chance of influx of water/ contaminant through porous egg shell. It should be wiped clean with 70 % alcohol before storage and before breaking it.

The forceps and hands should be wiped clean with 70 % alcohol before breaking the egg. The egg should be broken into two halves and the white of the egg should be drained into a beaker. The yolk is transferred to a clean and sterilized filter paper and rolled over the filter paper so as to remove the white. The yolk membrane is then punctured by sterile straw/ needle to drain yolk in a sterile measuring cylinder. Required quantity of yolk is prepared for reconstitution of the diluents.

7.2 Semen Processing for Cryopreservation

Semen processing for cryopreservation involves the following steps:

7.2.1 Semen Dilution

7.2.2 Filling and sealing of semen

7.2.3 Printing of the straws

7.2.4 Equilibration in Cold handling cabinet

7.2.1 Semen Dilution

Semen dilution is done for cryopreservation by packing 20-30, million progressively motile sperm per straw. Semen dilution should be undertaken in a laminar air flow taking all aseptic precautions. The semen sample and the extender should be at the same temperature at the time of semen dilution (34 °C). The semen dilution should preferably be done in a conical flask containing freshly prepared extender pre warmed to 34 °C. The collection tube containing the semen should be introduced into the tilted conical flask and the semen is gently poured into the extender directly. The semen and the extender are then gently mixed.

7.2.2 Filling and Sealing of Semen

The diluted semen is filled into straws using automatic filling and sealing machine under a laminar air flow. Machine with filling and sealing speed from 3,600 straws to 15,400 straws per hour are available. Disposable sterile tubing should be used for filling and sealing of semen. Laboratory end of the straw is sealed by ultrasonic vibrations and pressure leading to molecular changes in the straw plastic and consequent sealing. In the German minitube straw both the ends are sealed using plastic or steel balls. Approximately, 1-1.2 cm air space is kept near the laboratory end of the straw. This is done because the semen expands in volume after freezing and also to avoid contamination of semen while cutting the laboratory end of the straws at the time of insemination.

Alternatively manual filling can also be done with the help of vacuum pump, filling comb, rubber tube, straw clips and polyvinyl alcohol powder. The diluted semen is placed in the bubbler disc rested on a bubbler stand. The straws to be filled are clamped on a straw clip and the factory seal end is then fitted to a filling comb. The filling comb is attached to the vacuum pump with the help of rubber tubing. Filling is then done slowly by operating the vacuum pump and dipping the open end of the straw in the semen. Negative pressure helps the semen to be filled into the straws. Appearance of two distinct bands at the factory seal indicates the complete filling of the straws. Air space is created in all the filled straws by pressing the Open end/ laboratory end of the filled straws against the teeth of the bubbler. Sealing of the straws is done manually by polyvinyl alcohol (PVA) powder. PVA powder is spread on a clean glass petridish so as to form a layer of 4-5 mm. The open/ laboratory end of the straws are pressed against the powder so as form a seal of 4-5 mm at the laboratory end of the straw. Immediately after sealing, printing of relevant details on the straws should be done.

7.2.3 Printing of the Straws

Details like Bull no., Breed, Name of the laboratory, Batch no, Year etc. should be printed after filling of semen in the straw for proper identification. Automatic straw printing machine can be used for printing on the straw.

Straw holding clamp

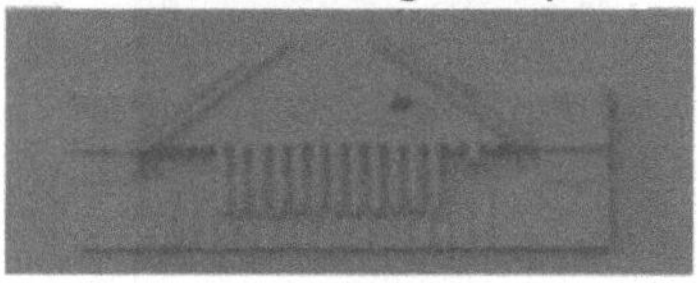

Air bubbler

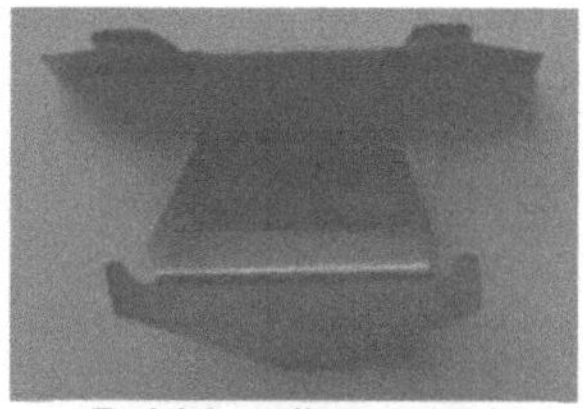

Bubbler disc stand

Bubbler disc

Filling comb

Bubbler disc assembly

Filling and sealing machine

Apparatus for manual and automated filling and sealing of semen

Immediately after printing straws are transferred in the cold handling cabinet and arranged in the freezing racks with the help of ramp which also facilitates easy counting.

7.2.4 Equilibration in Cold Handling Cabinet

Semen straws arranged in freezing racks are kept at 4 °C for 3-4 hours in cold handling cabinet. Equilibration period is the time between the addition of the cryoprotectant and initiation of freezing. The motility and metabolic activity of semen is decreased during this period and the spermatozoa become permeable to glycerol and an ionic and osmotic equilibrium is established with the extender. This also helps the spermatozoa to gain resistance against the freezing stress.

Chapter 8

Cryopreservation of Semen

Deep freezing or cryopreservation of semen refers to freezing and storage of semen at –79 °C using dry ice and alcohol or at -196 °C using LN2 as refrigerant. Cryopreservation of semen was first attempted by Parkes in 1945. But cryopreservation of semen received a boost by discovery of glycerol a cryoprotectant by Polge in 1949. Bull semen was first successfully frozen by Stewart in 1951. Cooling rate is a vital factor determining the post thaw survival of the sperm cells. Cell death occurs in rapid freezing due to formation of intracellular and possibly extracellular ice. In contrast slow cooling rates cause cell death due to diffusion of intracellular fluid into the extender, which then become increasingly hypertonic as the extender's water freezes causing intracellular osmotic damage as the cell dehydrates. Hence, the cooling rate needs to be balanced in a way that it is neither too fast to cause cell death due to cold shock nor too slow to result the same due to osmotic shock. Ultra rapid freezing gives excellent morphological preservation but with poor retention of functions on thawing due to large number of minute crystal formation. There are three major components in an ideal freezing curve, viz. the rate of decline of temperature from equilibration temperature (4 °C), the degree of super cooling which occurs below the extender's freezing point (-7 °C) and the efficiency with which the latent heat of fusion is absorbed by the nitrogen vapour input. Slow cooling is required between + 5 to - 5 °C and rapid cooling is required between -15 to -25 °C for bovine semen. Cooling rates below - 30 °C is not critical and sperm loss below - 30 °C were minimal.

8.1 Freezing in a Programmable Bio-Freezer

The equilibrated straws on a freezing rack are kept ready for loading. The liquid nitrogen container of the programmable freezer unit is filled with liquid nitrogen and the freezing unit is switched on and the already fed programme for freezing rate is selected. Run the programme and wait till the chamber temperature reaches 4 °C. At this stage the monitor will indicate "load samples". The samples are loaded in the freezing chamber. The programme will run its course and when the final temperature is reached, the monitor will indicate "remove samples" indicating that the freezing is completed. The straws are then collected in pre cooled goblets and immediately plunged into liquid nitrogen for further storage.

The following cooling rates give best results in cryopreservation of cattle and buffalo semen :

Temperature	Rate of Freezing (°C / minute)	
	Mini straw	Medium straw
+ 20 to + 4	-10	-10

Transfer freezing racks containing equilibrated semen from cold handling cabinet to freezer

+4 to - 12	-3	-5
-12 to -140	-60	-30

Transfer straws to pre cooled goblets and plunge immediately into LN 2 container

8.2 Manual Freezing

Horizontal liquid nitrogen vapour freezing technique is used for manual freezing of semen straws. This can be done in large liquid nitrogen storage containers or conventionally designed thermocool box.

Semen straws are placed in liquid nitrogen vapours 4 cm above the level of liquid nitrogen for 10-15 minutes till they reach a temperature of -130 °C to -150 °C and are then dipped in to liquid nitrogen directly for further storage at - 196 °C.

Straws being equilibrated in a cold handling cabinet

Programmable Bio Freezer

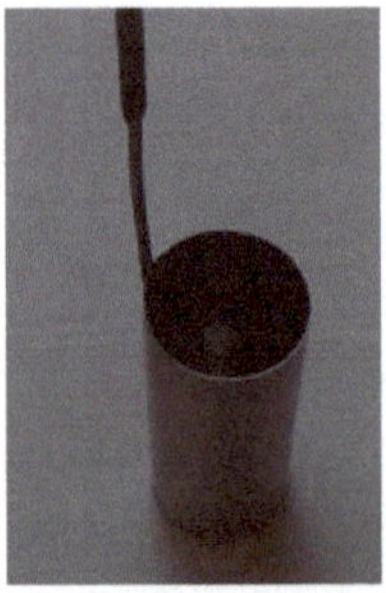

Canister

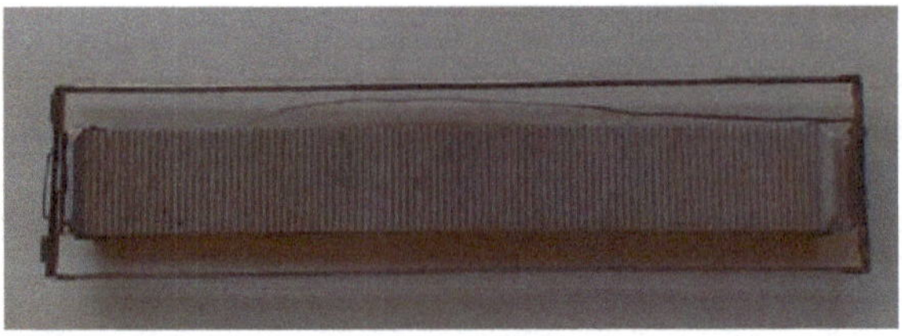

Freezing rack

Divider for goblet

Goblet

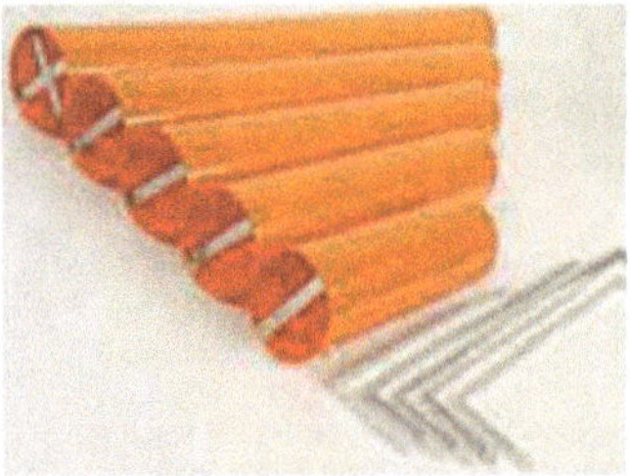

Canister with lifters

Freezing rack for 40 straws 0.5 ml

Distribution block for 40 straws 0.5 ml

Freezing rack for 175 straws 0.25 ml

Distribution block for 175 straws 0.25 ml

Chapter 9

Packing Systems for Frozen Semen

Early research in field of semen storage and delivery systems was conducted in USA. The straw technique for the storage of frozen semen was evolved by work of French, Danish and German workers whereas the pellet system of storage of semen was devised by Japanese scientists. The different packing systems of frozen semen are:

9.1 Ampoule method

9.2 Pellet method

9.3 Straw method

9.1 Ampoule Method

In this method the semen is placed in a glass ampoule and is subsequently sealed and frozen in liquid nitrogen. The ampoule is thawed just before insemination and is then cut open. The semen is withdrawn into glass catheter and used for AI. Since the ampoule is sealed there is no risk of contamination during storage. The relevant details can be marked on the ampoule. The major limitations of this method of semen storage involves lower freezability and fertility as the semen is stored in large volume in an ampoule, besides greater storage space as compared to straws and greater loss during handling at the time of thawing and insemination with glass catheter.

9.2 Pellet Method

Pellet method is a economical method and also occupies less storage space. Depressions are created on dry ice (solid CO2) and 0.1 to 0.2 ml of semen is deposited. This is left for 10 minutes and tablet/ pellets are then stored in goblets in liquid nitrogen at -196 °C. This pellet is dissolved in 0.9 ml of warm diluent and used like liquid semen. The major limitation of this method involves difficulty in identification, chances of contamination, moderate freezability and requirement of separate thawing solution. There is also high risk of breaking of pellet on handling with forceps.

9.3 Straw Method

History of introduction of plastic straws for storing liquid semen can be traced back to 1940 in Denmark. Adler in 1960 developed technique for freezing of semen in straws using liquid nitrogen vapours. Cassou in 1965 modified this technique for French straw made of polyvinyl chloride. Mini tub or German straw made of plastic, developed in West Germany, was introduced in 1972. U.S. or continental straw made of polypropylene were developed in USA. This method has got several advantages to other methods because of better freezability as the semen is processed in thin film, with greater area to volume ratio, enabling rapid heat exchange and better revival rates. The identification details can be easily printed over the straw and it requires less storage space requirement. Better hygienic standards can be met if automated filling is done. The French straws are nowadays widely used throughout the world. The French straw contains two ends. The manufacturer end is sealed by manufacturer by two cotton threads separated by a column of Polyvinyl alcohol powder (PVA) and laboratory end is left open. The laboratory end is sealed after semen is filled using automatic filling and sealing machine by ultrasonic vibration and pressure or manually by PVA.

Table: Dimension of various types of straws

Types of straws	Volume (ml)	Useful Volume (ml)	Length (mm)	Surface area (mm2)
French medium/ midi straw	0.50	0.48	133	1152
French mini straw	0.25	0.23	133	823
Continental / Minitube straw	0.25	-	65	555

Chapter 10

Good Laboratory Practices in Semen Collection and Processing

While collecting and processing semen the following points needs to be considered:

- Good quality Artificial Vagina (AV) should be used for semen collection. Strict hygiene should be maintained while collecting semen.
- Highly scientific and flawless collection technique should be used for semen collection.
- Neat semen should be maintained at close to body temperature.
- Semen should not come in contact with any contaminant such as lubricant, dust, infection etc. in the collection yard or in any stage of semen processing.
- Direct exposure of semen to sunlight or cool white fluorescent light should be avoided as it may cause chromosomal damage.
- Enclose all cool white light sources e.g. laminar flow hood, cold room etc. in gold or orange sleeves that absorb ultraviolet and low wavelength visible light (< 475 to 500 nm).

- Semen samples should not come in contact with temperature variations during collection and processing.
- The collection tube should be covered with aluminium foil and transferred to a water bath maintained at 35 °C.
- Immediately after collection, the collection tube containing semen should be transferred to a water bath maintained at 35 °C.
- Eating drinking or smoking should be strictly prohibited in semen collection yard or semen processing laboratory. Conversation should be discouraged during semen collection or processing.
- The use of latex surgical or examination gloves should be avoided during preparation of dilutor or semen processing due to potential toxic effect of powder residues on the glove surface. If the glove has to be used, it should be rinsed with sterile distilled water.
- The bare hands should be washed with soap and water and should be rinsed with 70 % alcohol solution before handling/ processing semen or preparing extender.
- Protocols for washing and sterilization of glassware, disposables and laboratory equipments should be carefully considered to avoid acquiring toxic elements during cleaning and sterilization.
- The work surface for semen handling and processing should be cleaned with 70 % ethanol before starting and after the end of the processing work, daily.
- Laboratory personnel and persons involved in semen collection must follow strict personal hygiene. Neat and clean laboratory coat, cap and mask should be put on before starting semen processing.

10.1 Good Laboratory Practices in Preparation of Semen Extender

- Laboratory personnel preparing semen extender should thoroughly wash and sterilize their hands and the working area in the laminar flow with 70 % alcohol.
- Only triple distilled water autoclaved at 15 psi for 20 minutes should be used for the preparation of semen extender.
- Only freshly prepared distilled water should be used for the preparation of semen extender because as little as 1 hour of storage in bottles may lead to leaching of new organics.

- Water fed to the distillation units should be tested for the concentration of heavy metals, calcium and iron as these elements are toxic to the sperm cells.
- Egg yolk used in the preparation of semen extender should ideally be specific pathogen free (SPF) as per international standards. But due to unavailability of such eggs, eggs from known sources/ farms should be used (to prevent mycoplasma infection).
- Never wash the eggs with water as it may lead to introduction of microbial contamination owing to porosity of egg shell.
- Wipe the eggs clean with 70 % ethanol. Only freshly procured eggs should be used for the preparation of semen extender. Eggs should not be stored in refrigerator for more than 2-3 days under any circumstances.
- Sterile whatman filter paper should be used for the separation egg yolk from the albumin.
- Excellent quality chemicals with highest purity should be used for the preparation of semen extender (AR Grade).
- pH of the semen extender should be checked regularly. Ideally it should have a pH of 7.0. Aberrant pH indicates serious error in the process of preparation of the extender. Merely adjustment of pH will not ensure good quality processed semen.
- Glycerol concentration in the semen extender should be regularly checked. (14 ml glycerol of 95 % purity and 86 ml distilled water gives 100 ml solution of 103.5 g weight).
- Concentration of glycerol in semen extender should be ideally 7 %. The glycerol used in the preparation of semen extender should have 99 % assay and should have almost nil content of heavy metals and free acids.

10.2 Good Laboratory Practices in Semen Dilution

- Correct dilution is important for quality assurance. Excessive dilution may lead to loss of intracellular constituents.
- Neat semen should be diluted immediately as delay in dilution may lead to decline in semen quality.
- If it is not possible dilute semen immediately, for time required for neat semen evaluation and calculation of dilution rate, it may

initially be diluted in 30 ml extender after recording neat semen volume.

- For precise dilution auto dispenser may be used.
- Conical flask should be used for dilution of semen. The centrifuge tube containing neat semen should be introduced into the conical flask and semen should be gently dropped into the extender.
- Neat semen should not come in contact of dry walls of the conical flask as it may induce bent tails.
- Immediately after dilution cooling (equilibration) of semen should be done.
- Bovine semen samples containing less than 500 million perm / ml and less than 60 % progressively motile spermatozoa should not be used for dilution and cryopreservation.

10.3 Good Laboratory Practices in Semen Packaging

- Good quality semen straws should be used for packaging of semen.
- Care should be taken to avoid contamination of semen during packaging/ filling in straws. Ideally the operation should be performed under a laminar flow.
- Extended semen should be mixed gently before filling to avoid the possibility of sperm cells settling at the bottom.
- Filling of semen straws can be accomplished with automatic filling and sealing machine. Alternatively, filling can be done manually with bubbler disc assembly and filling comb.
- Filled semen straws should be checked to ensure that they are properly sealed.
- Important information like bull no., batch and year of collection, name of semen station etc. should be printed on semen straws either with automatic printing machine or manually. It is better to use different color straws for packaging of semen of different breeds.
- In liquid semen approximately 10-15 million progressively motile spermatozoa should be packed in one insemination dose/ straw.
- In case of frozen semen approximately 20-30 million progressively motile spermatozoa should be packed in one insemination dose/ straw.

10.4 Good Laboratory Practices in Semen Packaging

- Spermatozoa are most susceptible to cold shock between 15 to 0 °C. Cold shock causes alteration in cell membrane permeability leading to damage of sperm cell and loss of sperm motility. Therefore, care must be taken to decide suitable cooling rate for bringing the sperm cell to equilibration temperature (4 °C). Cooling of bovine semen from the body temperature to 5 °C is carried out at a rate $<$ 10 °C per hour with incorporation of protective agents like egg yolk or milk/ glycerol.
- Ideally bovine semen should be equilibrated at 4 °C for 4 hours. During this period the osmotic adjustment of the sperm cells to addition of diluents and the process of maturation/ capacitation commence.
- Cooling is ideally done in 'cold handling cabinet'. Domestic refrigerators are not reliable for equilibration of semen.
- Semen straws should be placed in freezing racks before placing them in cold handling cabinet so that all the surfaces of individual straws are exposed to uniform cooling temperature.

Section C

Semen Evaluation

Chapter 11

Gross Evaluation of Semen

Gross evaluation is generally done to get a rough idea of the quality of semen and to assess the semen producing capacity of the bull. This also gives a fairly good idea about the reproductive health and any pathological conditions of the semen donor. The following are recorded under gross evaluation of semen:

11.1 Color and appearance of semen

11.2 Semen Volume

11.3 pH of semen

11.4 Semen consistency/ Semen density

11.5 Presence of foreign matter

11.6 Gross motility

11.1 Color and Appearance of Semen

The normal color of semen in different species is as mentioned below:

Species	Semen color
Bull	Milky white
Buffalo bull	Milky white to creamy white
Boar	
Pre- sperm fraction	Translucent
Sperm rich fraction	Whitish opaque
Post sperm fraction	Similar to Pre- sperm fraction but is less translucent
Stallion- Ejaculate consist of 6-9 jets resulting from contraction of urethra	
Pre- sperm fraction	Watery
Sperm rich fraction	Milky non viscous
Post sperm fraction	Highly Viscous
Last penile drip	Watery
Ram	Milky white or pale creamy
Buck	Grayish white or yellow
Camel	Whitish creamy to grayish translucent

Any abnormal color of the semen should be carefully observed and recorded and such semen samples should not be further processed for use in artificial insemination programme. The following deviations may be observed in color of semen owing to abnormalities mentioned against each of them:

Color of Semen	Possible cause
Red/Pink	Hemorrhage in the reproductive tract
Straw	Contamination with urine
Brown	Unhygienic collection or contamination with faeces
Brick red	Orchitis
Yellow	Contaminated with pus

11.2 Semen Volume

The volume of the semen is recorded immediately after semen collection preferably semen collection should be done in a graduated collection tube so that the volume can be immediately recorded after collection. Semen volume is affected by season, age, work load and effect of restraint/ teasing. Significant increase in semen volume is recorded in case of contamination of semen with urine (urospermia), blood (haemospermia) or pus (pyospermia). Lower semen volume may be observed in old age bulls, inadequate teasing and also in some clinical conditions like testicular rotation. The normal semen volume recorded in different species of animals is as under:

Species	Semen Volume (ml)
Bulls	
European Breeds	6-12
Indigenous Breeds	3-5
Buffalo Bulls	1.3-4.5
Boar	240-250
Stallion	60-100
Ram	0.5-2
Buck	0.5-2
Camel	2-10
Dog	2-10

11.3 Semen pH

The pH of semen is recorded immediately after collection by litmus paper or pH meter. Single electrode pH meter is the most suitable and most accurate method for determination of pH. Excellent semen samples will give pH of 6.4 - 6.5 in bulls and 6.8-7.2 in buffalo bulls whereas poor quality semen show pH towards neutrality. In conditions like incomplete ejaculation, extensive use of bulls and pathological conditions affecting testes/ seminiferous tubules/ ampullae, the pH of the semen will be increased (7 or more).

11.4 Osmolarity of Semen

The osmolarity of semen sample is measured by osmometer. Hypotonicity will cause swelling and deformation of the sperm cell especially the tail due to diffusion of water into the sperm cell. This may ultimately lead to rupture of the plasma membrane. Decreased osmolarity may be due to urospermia. Hypertonicity causes dehydration of sperm cells which adversely affects sperm functions.

11.5 Semen Consistency/ Semen Density

The semen should have homogenous consistency throughout the ejaculate and even the smallest visible difference indicates contamination like pus. Consistency of the semen should be assessed by slightly tilting the collection tube and observing the density of semen against the natural light.

Consistency	Expected sperm concentration
Watery	Less than 400 million spermatozoa / ml
Milky	400-800 million spermatozoa / ml
Thin creamy	800-1200 million spermatozoa / ml
Creamy	Above 1200 million spermatozoa / ml
Thick creamy	Above 2000 million spermatozoa / ml

Semen density is assessed similarly and graded as per Swedish workers as below:

Sperm density index	Expected sperm concentration
D	Less than 400 million spermatozoa / ml
DD	400-800 million spermatozoa / ml
DDD	800-1200 million spermatozoa / ml
DDDD	Above 1200 million spermatozoa / ml

11.6 Presence of Foreign Matter

The semen samples should be free from any foreign matter. If any foreign matter is present it indicates the unsuitability of semen for further processing. Pink or red color in semen indicates the presence of fresh blood due to small mechanical injuries of the glans penis or urethra, it may not be mixed up evenly in semen. Similarly brown color indicates decomposed blood originating from diseased condition in the deeper parts of the genital tract and it will often be evenly dispersed in semen. In freshly collected semen pus can be seen as flakes by tilting the collection tube. After some time pus settles down as yellowish, grayish or greenish sediment. Presence of pus indicates suppurative inflammation or infection in the genital tract.

11.7 Gross Motility

A drop of neat semen is placed on a clean glass slide and it is held above the source of light and on closer observation 'wave like' movements could be seen with naked eye. This type of gross motility could be seen only in exceptionally good semen sample.

Chapter 12

Microscopic Test for Evaluation of Semen

The following tests are used for microscopic evaluation of semen:

12.1 Sperm motility

12.2 Per cent Live sperm

12.3 Sperm concentration

12.1 Sperm Motility

The following two methods of estimation of sperm motility are followed in routine practice:

12.1.1 Mass activity

12.1.2 Progressive sperm motility

12.1.1 Mass Activity

Mass activity can be defined as 'en-mass activity of the spermatozoa in the semen, immediately after collection'. Mass activity is generally assessed by vigour of swirls and waves motion in undiluted semen under 20 X magnification and expressed on 0 to 5 scale. Mass activity is not very precise estimate of sperm motility but it is easily discernable characteristic of semen, which gives a rough estimate of the progressive motility of sperm in the semen sample and can be easily performed at AI centers. Though it is a subjective evaluation but it can give fairly good idea about semen quality and subsequent fertility. But the evaluation of mass activity alone should not be relied upon as in some species lower mass activity may not actually result in lower progressive sperm motility and viability. Some semen samples which have low mass activity can gain reasonably good progressive motility after dilution in appropriate extender. The mass activity differs significantly with bulls and seasons and was also found to be significantly positively correlated with the sperm concentration, progressive motility and live spermatozoa and has negative correlation with sperm abnormalities.

Procedure for Estimation of Mass Activity

Materials Required

1. Phase contrast microscope with stage biotherm
2. Water bath
3. Sugar tube
4. Sugar tube stand
5. Microslide

Mass activity of a semen sample is evaluated by placing a drop of neat semen over a micro slide, pre warmed at 37°C, and observing the type of movements of spermatozoa under 100 X magnification of a microscope fitted with satge biotherm (maintained at 37°C) and observing without a cover slip.

Observation

Observe under 100 X magnification and grade the motions on 0-5 scale as follows:

5 Very quick wave motions in the field. Only waves after the waves appear in the field. It is very difficult to trace when the eddies are formed and disintegrated. Swirls are created with mass movements of the spermatozoa.

4 The swirls and eddies are not comparatively so rapid as in Grade 5. The swirls can be observed to move towards extremities.

3 The swirls are slow and are scattered in the field. Individual movements of spermatozoa are discernable. Grossly, about 40-60 % of sperm cells are progressively motile and the rest are in weak motion.

2 Swirls are absent. Individual movements of spermatozoa is more evident in the field. Grossly, the observer finds not more than 40 % of the sperms in progressive motion. Other sperms are weak in motility. They show oscillating, undulating or circular movements. The non-progressive movements may also be characterized as 'throbing movements'

1 No wave motions are observed in the film, and only about 20 % of the spermatozoa have progressive movements. Rest of the spermatozoa show 'throbing movements'

0 The spermatozoa are not motile.

Inference: Semen samples with mass activity +2 or more are suitable for use in AI.

12.1.2 Progressive Sperm Motility

The physical activity of the sperm cells is necessary for sperm transport in the female reproductive tract and hence the estimate of progressive sperm motility can provide a fairly accurate prediction of semen quality. Progressive motility of sperm in a semen sample can be defined as the percent of spermatozoa moving in straight forward direction. This characteristic has long been a recognized estimate of the livability and fertilizing ability of the sperm cells as it bears a highly significant correlation with livability and fertilizing ability. The progressive sperm motility is significantly correlated with sperm concentration, per cent live sperm, cold shock resistance, freezability and fertility. It is significantly affected by individuals, breed and age of the bull and season. Progressive motility is also reported to decrease with the increase in dilution rate.

Procedure for Estimation of Progressive Sperm Motility

Materials Required

1. Phase contrast microscope with stage biotherm
2. Water bath

3. Sugar tube
4. Sugar tube stand
5. Microslide
6. Cover slips
7. Scissors

Thaw the semen straw at 37 °C for 30 seconds. Wipe the straw with tissue paper. Cut the laboratory end and collect the contents in clean sterile sugar tube. Alternatively, dilute neat semen in tris/ citrate buffer in the ratio of 1:10. Place a very small drop of frozen thawed/ fresh diluted semen on a clean greese free microslide. The microslide should be pre warmed at 37 ^{0}C. The semen drop is then covered with an 18 mm glass coverslip. The drop of semen should spread under the coverslip by itself. Preparation of the drop for motility examination is very crucial. Drastic variations in per cent progressively motile sperms are observed between drops prepared by skilled and unskilled examiner. At least 4-5 fields should be examined for assessment of progressive sperm motility. Count the number of progressively motile sperms and non motile sperms in each field and take the average. Avoid taking into consideration sperms near air bubbles or at the edge of the coverslip as it will lead to over or under estimation.

Observation

Observe the spermatozoa at 20 X phase contrast objectives to evaluate the samples for progressive sperm motility. At least 4-5 fields should be examined.

Inference

Fresh semen with less than 70 % progressively motile spermatozoa should not be frozen and frozen semen doses with less than 50 % progressively motile spermatozoa should not be used for AI.

12.2 Per Cent Live Sperm

While deciding the utility of a bull for artificial insemination it becomes important that its semen contains adequate number of fertile spermatozoa to reach the site of fertilization. The viability of spermatozoa is assessed by the percentage of live spermatozoa in the fresh or frozen semen. A thorough and quantifiable evaluation of semen along with percent live spermatozoa is desirable and may be achieved by differential staining technique using Eosin-Nigrosin stain. It is based on the principle that the membrane permeability of dead sperm increases as it becomes permeable to stains and appears colored whereas a live spermatozoa will not take stain.

The live percent of spermatozoa in a semen sample is affected by species and bulls within the same species, ejaculate frequency, age of bull, sex libido, nutritional status and season. Per cent live sperm bear a significant positive correlation with sperm motility and fertilizing ability.

Estimation of Per Cent Live

Materials Required

1. Water bath
2. Microscope with 100 X oil immersion objective
3. Immersion oil (cedar wood oil)
4. Microslide
5. Sugar tube
6. Scissors

Chemicals/ Reagents Required

Eosin-nigrosin stain

Any one of the following Eosin-nigrosin stain preparations can be used:

Hancock's (1951) Stain

Eosin Y	5.00 g
Nigrosin	30.00 g
Distilled water	300.00 ml

Dissolve nigrosin in distilled water while stirring and heating until it is dissolved. Later dissolve eosin in the nigrosin solution in a similar manner. Do not boil. Campbell et al (1960) Stain

Nigrosin Solution	150.00 ml
Eosin Y (GT Gurr)	5.00 g
Stock Buffer Solution	30.00 ml
Stock Glucose Solution	30.00 ml
Distilled Water up to	300.00 ml

Dissolve by stirring and heating over a hot plate. Do not boil. Filter and store at 4-5°C.

Smear Preparation and Staining Procedure

1. Thaw a straw of frozen semen at 37 °C for 30 seconds in a water bath.
2. Maintain a small aliquot of the eosin nigrosin stain at the same temperature for at least 5 minutes before staining.
3. Place 1-2 drops of stain and a very small drop of neat semen (approx. 1/5th of the stain) at one end of a clean microslide. A comparatively larger drop is taken for frozen thawed semen.
4. The drop of stain and semen is mixed with the help of the edge of another microslide without crushing or injuring the sperm cells. French medium straws may also be used for this purpose.
5. Allow this mixture to stand for 1-2 minutes.
6. Draw a smear of this mixture with the help of another slide with smooth edge. The second slide used for preparing the smear should be drawn smoothly by holding it at an angle of 45° and ensuring that the sperm stain mixture is in the inner angle of the two slides. Glass slide with rough edges should not be used for drawing smears.
7. Dry the smear quickly by blowing hot air with blower or hair dryer or by keeping at warm stage maintained at 37°C. In no case should the smear be dried over a flame.

Observation

The slides should be observed under oil immersion at 1000 X magnification to count the number of live sperm cells. Atleast 200 cells should be counted. The live sperm appears white against dark background of nigrosin whereas the dead sperm takes the eosin stain and appears pink.

Calculate per cent live as per the following equation:

$$\text{Per cent live sperm} = \frac{\text{Total nos. of live sperm counted}}{\text{Total nos. of sperm cells counted}} \times 100$$

Inference

Semen sample containing less than 70 % live sperm cells in case of neat semen and 55 % in case of frozen thawed semen should not be used for AI.

Estimation of progressive sperm motility

Placing semen on the micro slide

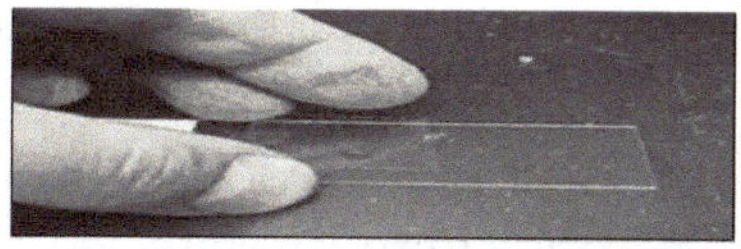

Putting cover slip on the semen drop

Estimation of sperm concentration

Neauber's counting chamber

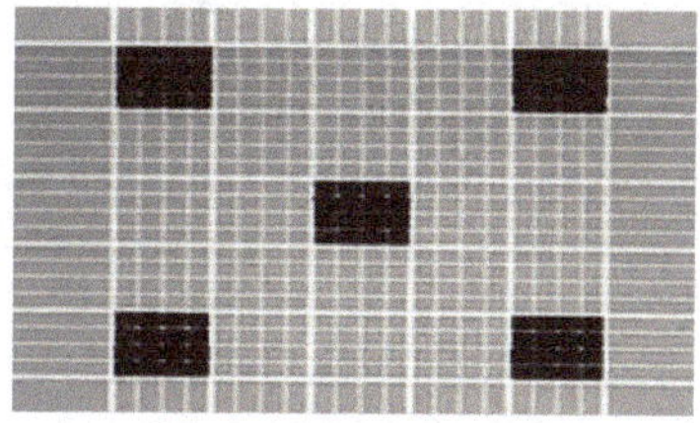

Squares where sperms are counted are indicated in dark shed

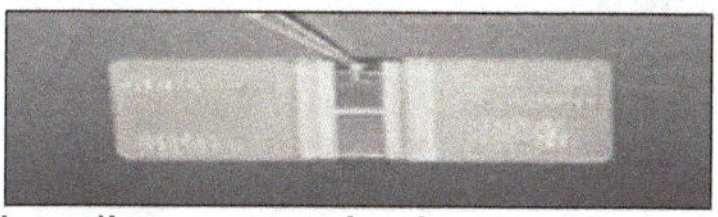

Loading a neauber's counting chamber

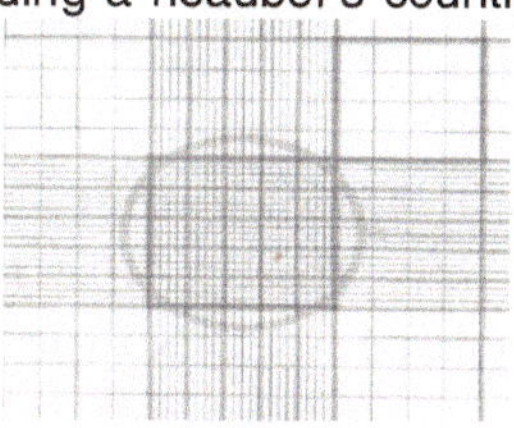

Neauber's counting chamber

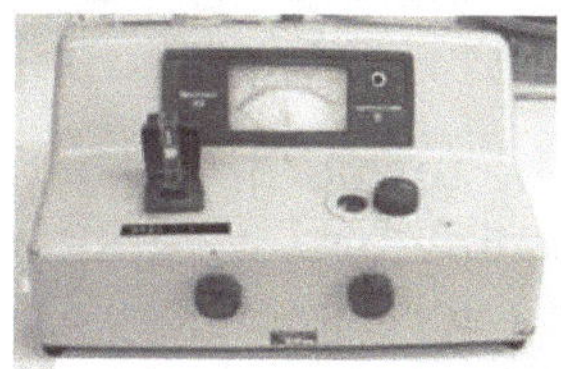

A Spectrophotometer

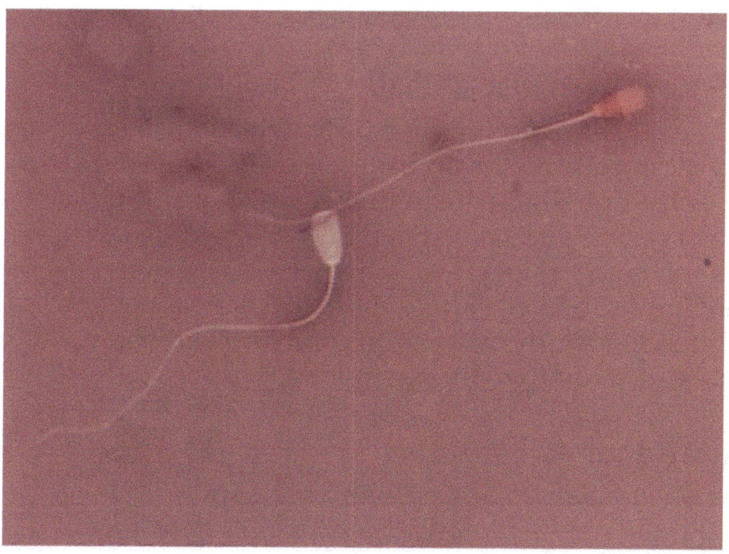

Live & dead sperm

12.3 Sperm Concentration

Sperm concentration is an important parameter determining the semen quality and has direct bearing on the number of semen doses to be produced for AI. It is one of the most important parameters to assess, once viability has been assessed. It gives a fairly good idea about reproductive capability of the bull and is also of prime importance in deciding the number of liquid / frozen semen doses that can be produced from a single ejaculate. Therefore, sperm concentration significantly affects the potential value of an ejaculate.

Sperm concentration is a highly variable trait, which is affected by breed and age of bull, season, number of false mounts, period of sexual rest, successive ejaculations and climate.

The sperm concentration can be evaluated by only one the following methods:

12.3.1 Haemocytometer method

12.3.2 Photoelectric colorimeter method

The former is by far the most economic option but is more time consuming and gives comparatively less consistent results. But most of the colorimeters are also calibrated on the basis of haemocytometer counts and depends on this method only.

One of the most recent method with high accuracy and repeatability is use of Computer assisted semen analyzer (CASA).

12.3.1 Haemocytometer Method

The Neauber's counting chamber originally used for counting the blood cells can be used for determining the sperm concentration also. It is accurate but time consuming method especially when large numbers of samples are to be examined as it require considerable skill and care. The advantage it has over spectrophotometer method is that the identification of individual sperm is possible and sperm cells can be differentiated from the debris and hence over estimation due to debris present in sample can be avoided.

Materials Required

1. Hemocytometer chamber with coverslip
2. Test tube
3. Test tube stand

4. Capillary tubes
5. 0.1 and 1.0 ml pipette or micropipette
6. Scissors

Reagents/ Chemicals Required

Eosin diluting fluid

Eosin diluting fluid	
Eosin Y water soluble	0.05 g
Sodium Chloride	1.00 g
Formaldehyde	1.00 ml
Distilled water up to	100 ml

Estimation Procedure for Estimation of Sperm Concentration

1. Take 9.9 ml (Tube 1) and 9 ml (Tube 2) Eosin diluting fluid in two test tubes.
2. Add 0.1 ml fresh or frozen thawed semen to Tube 1.
3. Mix gently.
4. Transfer 1 ml of the contents of Tube 1 to Tube 2.
5. Mix gently.
6. Fix coverslip on Neauber's counting chamber and charge the chamber with the contents of Tube 2 with the help of a capillary tube/ micropipette. Focus the chamber under 200 X or 400 X magnification of a microscope.
7. Count the total number of sperms in four corner squares (Top left, top right, bottom left and bottom right) and one central square.
8. Sperm cell should be counted only if whole head of the sperm is inside one of the above mentioned squares. If head is completely outside the squares, the cell should not be counted. The investigator may, however, adopt a uniform and impartial standard in case a part of the head is inside the square to be counted.

Observation

Count the total number of sperm cells in four corner squares (Top left, top right, bottom left and bottom right) and one central square under 200 or 400 X magnification.

Let the number of sperms counted as above be N.

Sperm concentration per ml of semen = $NX50X10^6$

Calculation of Sperm Concentration

A Neaubers counting chamber contains 25 big chambers. Each big chamber, in turn, contains 16 smaller chambers/ squares.

Total number of small squares = 25 X 16 small squares

= 400 small squares

Total area of 400 small squares = $1\ mm^2$

Total number of small squares counted = 5 X 16

= 80 small squares

Area of 80 small squares = $\frac{1 \times 80}{400}\ mm^2$

= $1/5\ mm^2$

Depth = 1/ 10 mm

Volume of 80 small squares = $1/10 \times 1/5\ mm^3$

= $1/50\ mm^3$

If N number of spermatozoa are counted in 80 small squares

i.e. Number of sperm cells in $1/50\ mm^3$ volume = N

Number of sperm cells in 1 mm3 volume of semen = N X 50

Number of sperm cells in 1 ml volume of semen = N X 50 X 1000

= 50 X N X 1000

Total sperms per ml of semen = 50 X N X 1000 X 1000
(Dilution-1000 times in eosin diluting fluid)

= $50 \times N \times 10^6$

= $50 \times N \times \text{million}\ ml^{-1}$

12.3.2 Photoelectric colorimeter method

Materials Required

1. Spectrophotometer with Cuevettes
2. Test tube
3. Test tube stand

4. 0.1 and 5.0 ml pipette or micropipette
5. Scissors

Estimation Procedure

1. Use 720 N filter.
2. Take the 2.5 ml sodium citrate buffer (2.9 %) in the special colorimeter test tube. Wipe clean with tissue paper and fix in the slot of instrument and switch on the instrument. If needed, adjust the knob and set the needle to 100 percent light transmittance.
3. Add 0.1 ml neat semen and mix gently. Place the cuvette in spectrophotometer and record the reading. Translate the reading into sperm concentration with the help of a standard graph prepared with the help of haemocytometer.

Inference

The frozen semen doses should ideally contain approximately 20 million progressively motile sperm per dose for optimum fertility.

Range of normal sperm concentration in neat semen of various species (million ml-1)

Species	Concentration
Cattle	800-2000
Buffalo	300-1200
Sheep	2000-3000
Goat	2000-3000
Pig	200-300
Horse	150-300
Dog	100-250
Poultry	3000-7000
Human	50-150

Chapter 13

Sperm Morphology

13.1 The Normal Sperm Morphology

A mature spermatozoa is an elongated cell consisting of a flattened head containing the nucleus, a neck and a tail containing the apparatus necessary for cell motility. The sperm cell consists of the following parts covered by the plasmalemma or plasma membrane:

13.1.1 Head

13.1.2 Neck

13.1.3 Tail / flagellum - The sperm tail consist of

a. Mid piece
b. Main piece / Principal piece
c. End piece

Sperm Plasma Membrane

The sperm plasma membrane is three layered membrane. It lies in close contact with the underlying cytoplasm and attached at three places i.e. over anterior outline of the sperm head, around the base of the head, in circle around the distal end of the midpiece (Jensen's ring). This attachment causes creation of three compartments with probably no communication with each other. The sperm plasma membrane is invisible under normal light or phase contrast microscopy but can be visualized after addition of substances with light refractory index e.g. bovine plasma albumin or synthetic polysaccharides. It is very easily permeable to water but there is low permeability for inorganic cations (Na, K, Ca etc.) and even lower permeability for anions (NO3, Cl). Exposure of sperm cells to hypertonic media will cause the water to go from sperm to the surrounding media leading to shrinkage of sperm cell whereas exposure of sperm cells to hypotonic media will induce influx of water causing swelling of the sperm cell. The more water that goes into the tail the more it changes its shape and becomes rounded to give room for increasing amount of water. As the length of the tail decreases as it becomes round, the tail fibrils have less space and tail bends and coils within the membrane vesicle. Tail fibrils are not damaged initially and will continue to contract till the critical volume of bull sperm is 3.8-3.9 times of its volume in an isotonic solution. The swelling and change in shape of the spermatozoa are momentary processes and can be corrected if hypotonicity of the media is corrected before the sperm cells reach the critical volume for rupture.

Sperm membrane is responsible for maintenance of the intracellular biochemical environment. Damaged plasma membrane is responsible for the leakage of electrolytes and organic substances such as enzymes and may also lead to damage of sperm cells by foreign substances from the surrounding media.

13.1.1 The Sperm Head

Nucleus is the largest part of the sperm head. It is oval, flattened and consists of densely packed chromatin surrounded by nuclear membrane. Chromatin consists of deoxyribonucleic acid (DNA) bound to the nuclear protein called sperm protamines. This appears as homogenous, compact mass on electron microscopy. This specific appearance of the chromatin is due to condensation, which starts in the spermatid and continues through the upper part of the epididymis. This condensation depends upon chemical changes of the nuclear protein and stronger binding of DNA to this protein. Small vacuoles like cavities are observed in this compact homogenous nuclear material on electron microscopy. Sperm nucleus gives a clear contrast at positive phase contrast microscopy. Nucleus can be stained with alkaline

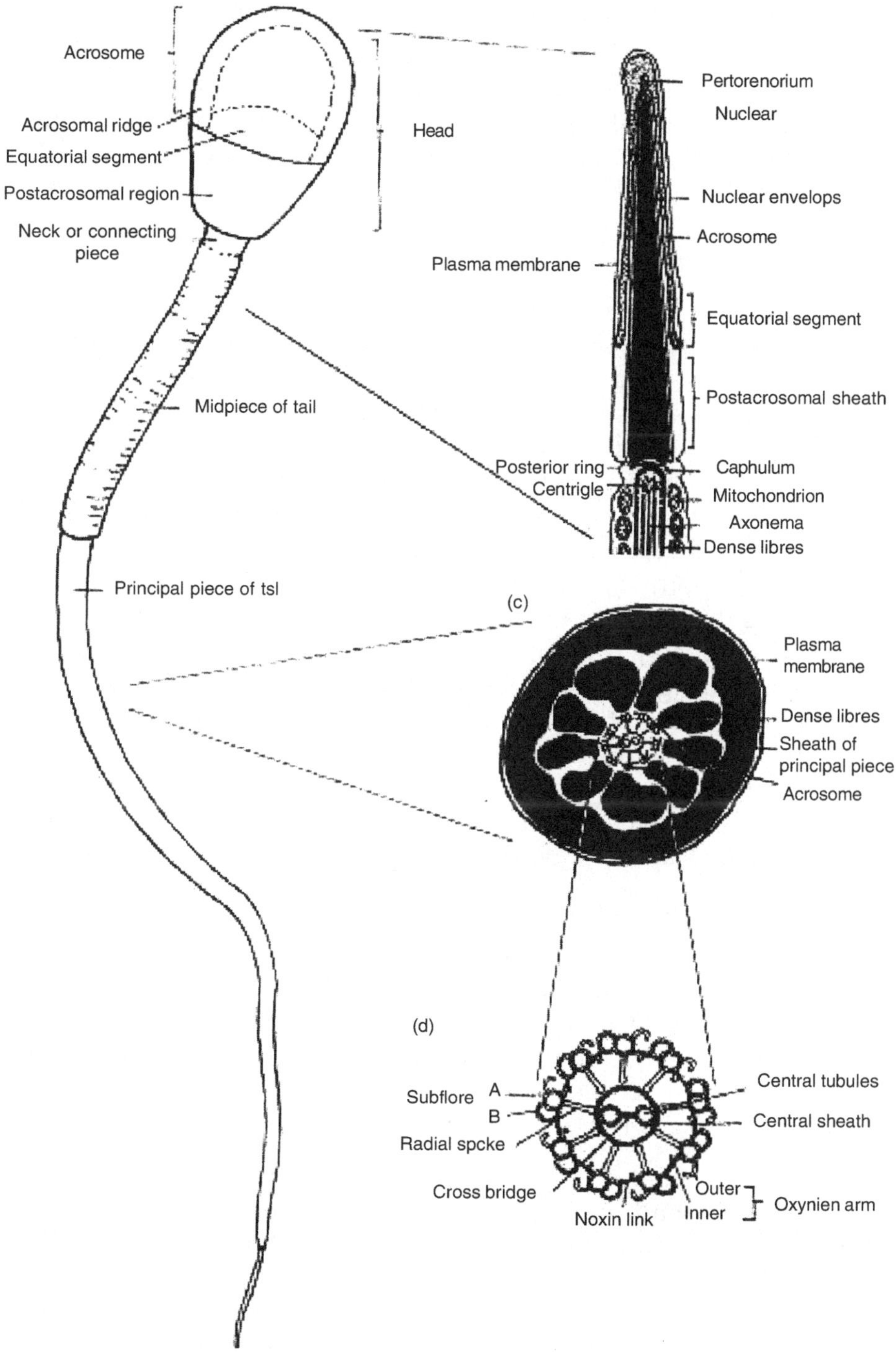

Normal sperm morpholoty and ultrastructure of bovine spermatozoa nuclear stain e.g. hematocyclene. Sperm DNA can be specifically stained by Feulgen stain. Quantification of sperm DNA can be done by light absorption in ultraviolet light. The chromosome number of the sperm nucleus is haploid or half of the somatic cells of the same species which results due to meiotic cell division that occur during sperm formation.

Acrosome appears like a cap like structure over the anterior part of the sperm head covering up to 75 – 80 % of the surface. It is covered by a double layered membranous sac (inner and outer acrosomal membrane) layered over the nucleus during the last stages of sperm formation. Between the membranes acrosomal contents are present which consist of acrosin, hyaluronidase and other hydrolytic enzymes which facilitates sperm penetration into the egg. Acrosome is thicker on one side along the anterior outline and partly down the edges of the head, forming apical ridge. Equatorial segment of the acrosome is crescent shaped segment in posterior part of the acrosome having a curved limit pointing forward. Equatorial segment is important because it is this part of the sperm cell, along with the anterior portion of the post acrosomal region, which initially fuses with the oocyte membrane during the fertilization process. This area can be clearly observed in phase contrast microscope and can be stained with Giemsa. Acrosome is a very delicate structure. In fixed semen samples stored at room temperature or refrigerator, the acrosome may be swollen and the acrosomal membrane might rupture. Therefore, examination for acrosomal integrity should be done on very fresh semen samples.

13.1.2 The Sperm Neck

Sperm neck is a morphologically complicated structure that connects the sperm with tail at a point called implantation. As a rule implantation is located centrally in the base of the head. But in some cases the implantation can be displaced to one side, so called abaxial tail implantation (Paraxial implantation in German and Excentric implantation in English).

Sperm cells contain centrioles which participate in the cell division by spindle formation. During transformation from spermatid to spermatozoa, the centrioles makes contact with the spermatid nucleus and lead to the formation of the implantation fossa in the nucleus. The formation of the tail is initiated from the distal centriole whereas the proximal centriole via the basal plate is fastened in the implantation fossa and is surrounded by segmental implantation plate. Further down the tail is strengthened by 9 course fibrils which forms a cylinder around the central fibers. The course fibers continue along the tail distally through the midpiece and partly down the tail proper as 9 separate fibers.

13.1.3 The Sperm Tail

The central stem of sperm tail consist of 2 central tubuli and a ring of 9 peripheral fibril doublets, of which one is tubular and the other one is compact. The compact fibril has got arms directed to the nearest doublet and is also connected to the central fibril by spokes. Outside the peripheral doublets is a ring of 9 course fibers running from the neck and down most of the main piece. In transverse section three of the course fibers are larger than the other ones and are numbered 1,5 and 6.

The midpiece is thicker as compared to the main piece/ principal piece due to presence of a mitochondrial helix made up of chains of elongated mitochondria. Below the mid piece the main piece of the tail is surrounded by a fibrous sheath. In light microscopy only the thicker mid piece and the thin end piece can be differentiated from the main piece of the tail.

Shape and Size of the Sperm

The sperm head of bulls is largest and stallion is smallest. The shape of the bull sperm head is thin tongue like plate with an evenly rounded anterior outline. The acrosome is clearly visible. Abaxial attachment of the tail occurs normally in stallion and dog sperm.

Biometry of cattle and Buffalo spermatozoa

Category	Sperm Dimensions (μ)	
	Cattle	Buffalo
Head		
Length	9.15	7.47
Breadth	4.25	5.52
Midpiece		
Length	14.84	11.76
Breadth	0.67	-
Main piece		
Length	45-50	43.1
Breadth	0.51	-

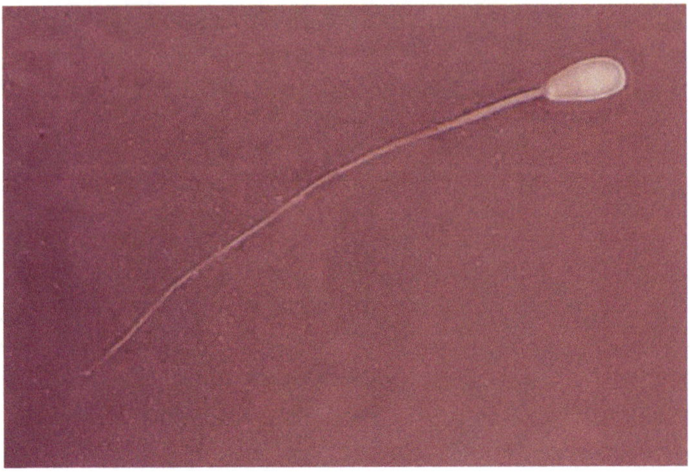

Normal sperm

13.2 Abnormalities in Sperm Morphology

The sperm abnormalities are assessed as it has significant negative correlation with viability, motility, in vitro fertility and also in vivo fertility. Roberts (1982) suggested that a good semen sample should not contain more than 20% abnormal sperm whereas the American society for theriogenology considered ejaculates containing less than 30 % morphologically abnormal spermatozoa are fit for insemination.

Sperm abnormalities may be hereditary in nature e.g. abaxial attachment of the midpiece to the sperm head while those of protoplasmic droplets and tailpiece are variable from season to season or may depend upon the age of bull and frequency of collection.

Sperm abnormalities may be classified as head, body and tail defects (Roberts, 1982). Lagerlof and Blom (1950) classified these three anatomical groups of sperm abnormalities in two further groups as primary sperm abnormalities that occur due to disorders in the seminiferous or germinal epithelium, and secondary sperm abnormalities that occur after the sperms have left the germinal epithelium. Blom (1977) classified sperm abnormalities into two groups as Major and Minor sperm defects. Recently Saacke had classified morphological abnormalities leading to infertility or reduced fertility that can be minimized or eliminated by adjusting the quantity of sperm as 'compensable' or 'non-compensable'. Compensable defects are those sperm defects wherein the normal fertility can be maintained by increasing the number of spermatozoa whereas the non compensable defects are those defects wherein the normal fertility cannot be restored even after increasing the number of sperms in the insemination dose.

13.2.1 Abnormalities of Sperm Head

The following types of abnormalities are frequently noticed in the sperm head:

Free or Loose or Detached Heads

This defect can be categorized into free normal head; free abnormal heads and decapitated sperm defect. Incidence of free normal head in small percentage is common in normal fertile bulls. Detached sperm head may be found in greater numbers in smears prepared by the beginners or inexperienced workers with rough hand. Higher percent of free heads may also be found in semen of bulls suffering of partial testicular hypoplasia, testicular degeneration or seminal vesculitis syndrome. Toxaemia and laminitis that compels a bull to rest most of the time lead to a very high incidence (91 percent) of free heads. Separation between head and tail occurs in the caput epididymis due to a narrow implantation groove and abnormal structure of the implantation plates. Decapitated sperm defect is characterized by separation into loose heads and tails of nearly 80-100 % of the ejaculated sperm, energetic movement of detached tail and a proximal bending of the free tails around the cytoplasmic droplets which looks like microheads.

Knobbed Sperm

This defect is mainly found in Holstein Friesian and Charolais breed. Knobbed sperm defect is characterized by a flattening or indentation of the apex of the acrosome with about 20-50 percent of spermatozoa showing a refractile bead below the indentation on light microscopy. A high incidence of knobbed acrosome defect is associated with sterility while a low incidence is related to infertility. The knobbed sperm defect could be hereditary and linked with recessive gene.

Detached Acrosome

The acrosome contains enzymes vital for penetration of spermatozoa into the Zona pellucida. During fertilization the acrosomal cap (galea capitis) undergoes changes in biochemical composition and ultra structure. Acrosomal abnormalities such as the "knobbed acrosome defect" lead to infertility or sterility. Ruffling, swelling, rupture, missing part (incomplete acrosome) and loss of acrosomal membranes are some of the other common aberrations in acrosome integrity. Ruffled acrosomes are characterized by a wrinkled appearance due to breakdown of the sperm cell membrane and acrosomal membranes and incomplete acrosomes by piece or pieces missing along the margin. These acrosomal abnormalities may be hereditary and may be involved in infertility.

Diadem Defect

This defect is characterized by appearance of a string of 5 to 6 beads, placed like a necklace around the sperm head at the level of acrosome-post acrosomal sheath junction i.e. at the level of anterior border of post-nuclear cap; hence the name diadem defect. This defect can be identified with eosin-nigrosin stain / rose bengal stain. A high incidence of the diadem defect is associated with disease condition or stress.

Abnormal DNA Condensation

Abnormal DNA condensation is characterized by a coarse or fine grained clumping in sperm nuclei stained with Feulgen's technique. Nuclear vacuolation is also one type of abnormal DNA condensation. This defect cause adverse effect on fertility. Trauma to the scrotum and testicles and an abnormal genetic constitution are predisposing factors for abnormal DNA condensation. Occasionally, a few spermatozoa appear as ghost cells with weakly staining nuclear chromatin. At times they do not stain at all by Fuelgen's staining.

Narrow and Pear Shaped Heads

These are the most common abnormalities in the shapes of head. In a narrow or tapered head, the entire nucleus is narrow and appears elongated. Spermatozoa with head narrow at the base differ from tapered heads. A pear shaped (pyriform) head is characterized by a rounded acrosomal and a narrow post-acrosomal region. Barth and Oko (1984) have categorized these head defects as the same as both forms (narrow and pear) can occur in the same semen with one form being predominant. Also, often a small to very large proportion of tapered or pyriform heads are affected by vacuolation. It is often difficult to differentiate slightly tapered sperm from normal sperm.

Despite motility and normal acrosomal morphology, the presence of a sizeable proportion of spermatozoa with pear-shaped heads leads to reduced fertility. Disturbed spermatogenesis may be due to testicular temperature for endocrine regulation of testicular functions, may lead to presence of pear shaped spermatozoa.

Abnormal Contour

A gross abnormality in the nucleus, like vacuoles spread throughout the nucleus and acrosomal cap, results in sperm heads with abnormal contours. Such sperm usually look glassy in eosin-nigrosin stained smears, presumably due to an abnormal DNA condensation. The condition is associated with the presence of other morphological defects in the semen from affected bulls. Usually such semen has poor motility and freezability.

Small and Giant Heads

Macrocephalic (giant head) and microcephalic (small) head represent sperm heads which are larger and smaller, respectively than the normal sperm heads. Though this abnormality is frequently observed in dairy bulls but its impact on fertility has not been established.

Rolled Head-Nuclear Crest-Giant Head Syndrome

In this defect the edges of sperm head are found curved to varying degrees giving rise to shapes varying from a slight curve to a tube with the curvature being present along the entire length of the head. Rolled heads may be a deviated form of giant head. Based on cytophotometric studies, it was observed that the DNA content of large heads was double as compared to the normal heads and were, therefore, diploid and suggested that such nuclear defects in spermatozoa could cause reduced fertility.

Double Forms

Occasionally, sperms cells with double or triple heads are also found but the incidence of such defects is very low.

13.2.2 Abnormalities of the Sperm Midpiece and Tail

Blom (1972) has characterized these defects under major and minor category as listed below:

	Major defects	Minor defects
i.	Corkscrew defect	i. Simple bent tail
ii.	Dag defect	ii. Abaxial implantation
iii.	Pseudodroplet defect	iii. Terminally coiled tail
iv.	Tail stump defect	iv. Distal droplets
v.	Other midpiece defects	
vi.	Proximal droplets	

Corkscrew Defect

In this defect the midpiece looks like a corkscrew due to irregular distribution of mitochondria along the mitochondrial sheath on light microscopy. The abnormality is usually observed in aged bulls in several exotic breeds. The affected spermatozoa are always dead and often have residual proximal cytoplasmic droplet. This defect is usually accompanied by other major defects and even a low incidence (5-10 percent) of corkscrew sperm is associated with reduced motility and fertility. The occurrence of this defect was associated with a progressive testicular degeneration. Recovery seems to be rare.

The Dag Defect

This defect was first reported in a Danish jersey bull named "Dag" by Blom. The defect is characterized by a folding and coiling of the midpiece associated with a fracturing of axonem at the midpiece with a disrupted mitochondrial arrangement. This defect seems to occur during passage of sperm through caput epididymis and the incidence in cauda epididymis is similar to that found in the ejaculate. Such defect was also seen in the semen of HF bulls suffering from testicular degeneration due to thermal stress in summer. The condition has also been found in buffalo bulls suffering from disturbed spermatogenesis.

Pseudo Droplet Defect

The pseudo droplet defect is another type of rare midpiece abnormality characterized by a rounded or an elongated thickening in the midpiece, often associated with a bent or a fracture at the same site and thus forming an angle. The thickened area may be confused with a dislocated cytoplasmic droplet, but an examination of eosin-nigrosin or India ink stained smear usually reveals vacuoles and granules within the thickened area. In Blom's study, the frequency of defect appeared to increase with age with simultaneous decrease in the sperm motility and fertility. This defect can be mistaken for a protoplasmic droplet. Fertility is affected due to this defect. It has been observed in HF bulls.

Tail Stump Defect

The "tail stump defect" has been named by Blom and has been observed in many breeds of cattle and several other species, but with a low incidence. The affected bulls usually have oligospermia associated with lower percent of motile spermatozoa because many spermatozoa lack a functional tail. In fact, on light microscopy, the condition may be indistinguishable from a high incidence of loose heads. Under higher magnification most loose heads are found to possess short (2-3 micron long) tail rudiments usually veiled by a cytoplasmic droplet-like rounded body. The defect is often associated with other defects like pyriform heads and Dag - like midpiece defects but loose sperm tails are seldom seen. The defect appears to be heritable and the affected bulls are usually sterile. Recovery seems unlikely as the incidence increases with age.

Simple Bent Tail

The most typical appearance in this defect is the bending of distal region of the midpiece in the shape of the alphabet "J" or a shepherd's crook. It is the most common aberration in spermiogram of bulls of all breeds. The incidence of defect may be upto 25 percent in the semen of normal fertile bulls. However, compared to the Dag defect, in this case the mitochondria is smooth and complete.

Bowed Midpiece

Very often a high proportion of spermatozoa show a rainbow or a wide U-shaped midpiece without any characteristic fold like the simple bent tail. In most cases the bent is an indication that sperm has all but lost its motility. This condition can also be differentiated from artefact induced by a faulty semen smear preparation as several spermatozoa with bowed midpiece in an eosin nigrosin stained smear will not show this condition in wet preparation, if the condition is an artefact.

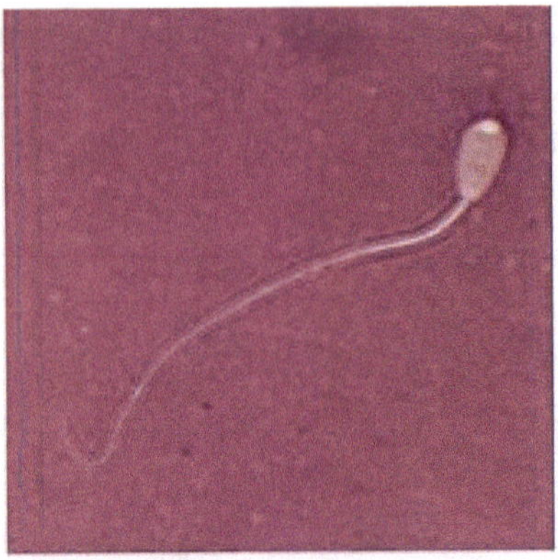

Knobbed acrosome

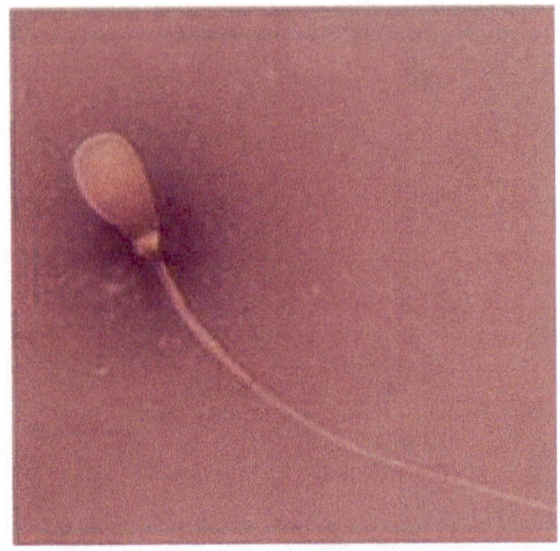

Narrrow at base

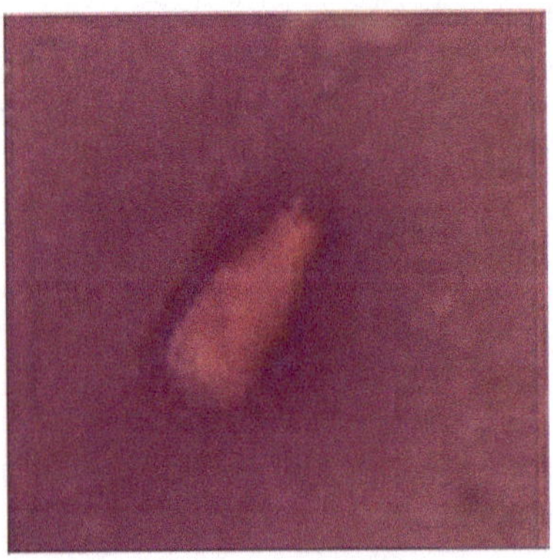

Free pathological head

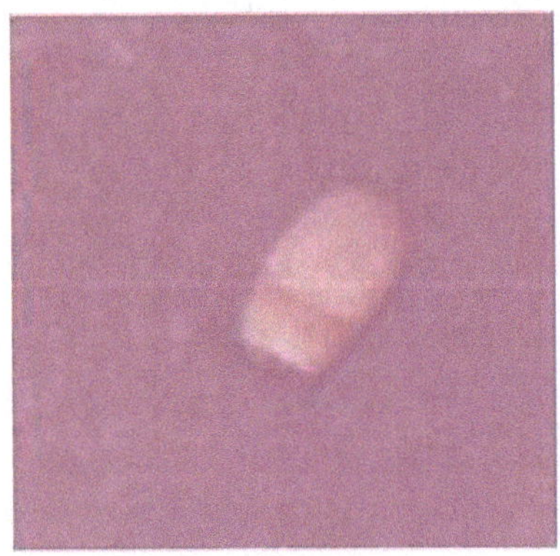

Free normal head

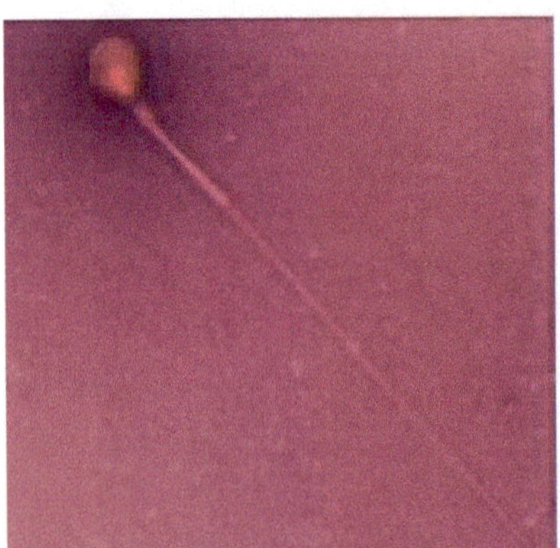

Microcephalic (Spherical)

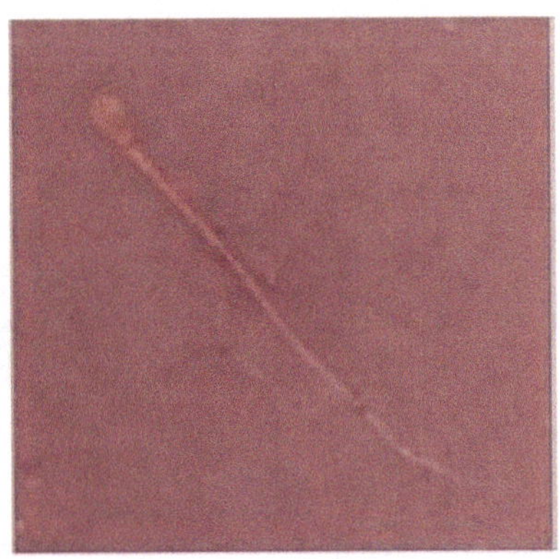

Microcephalic

Abnormalities of bull sperm head

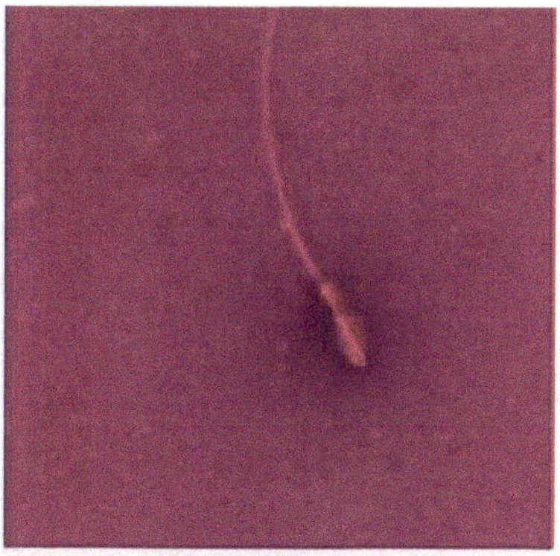

Asymmetrical head

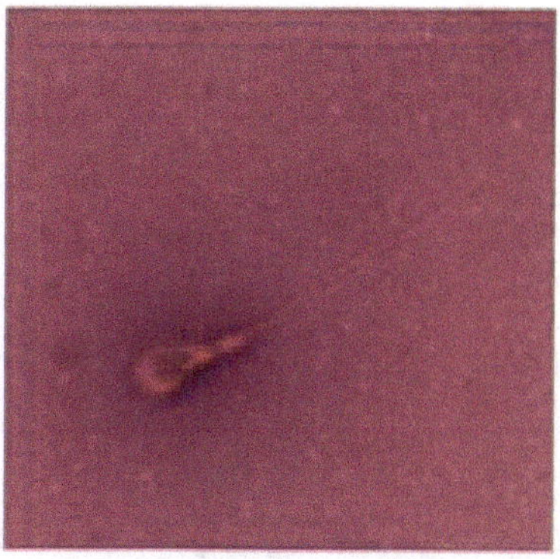

Asymmetrical head

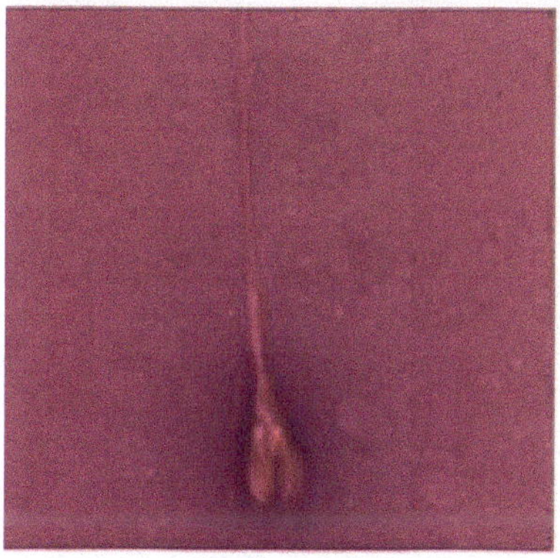

Asymmetrical head

Asymmetrical head

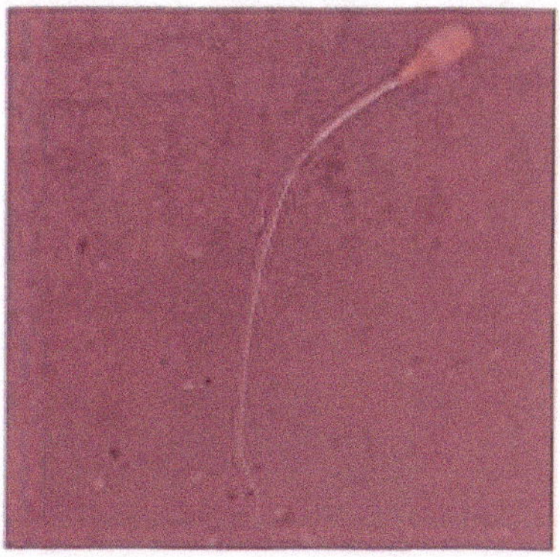

Narrow head

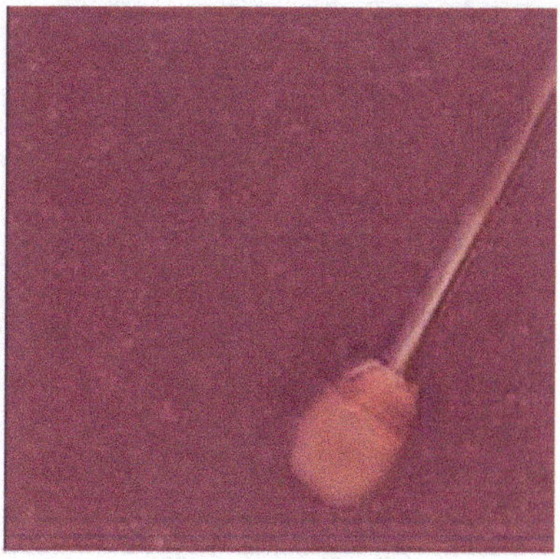

Macrocephalic

Abnormalities of bull sperm head

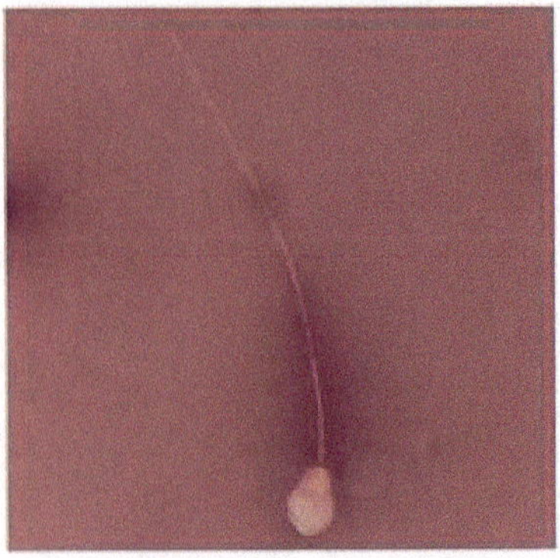

Pear shaped head

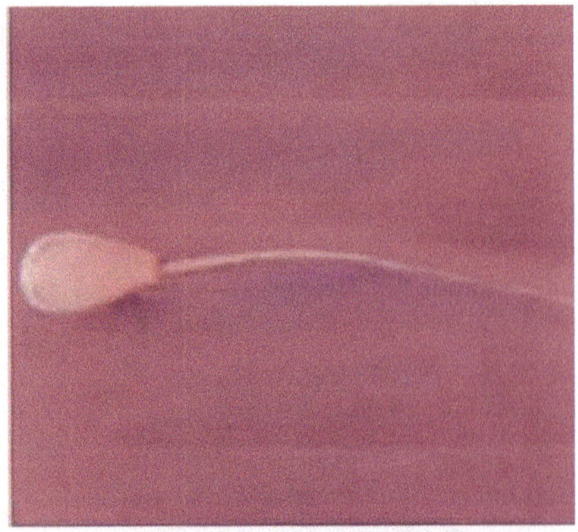

Pear shaped head

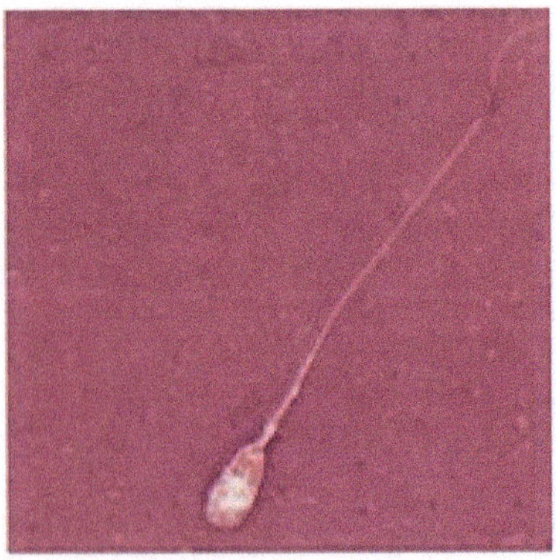

Abnormal contour

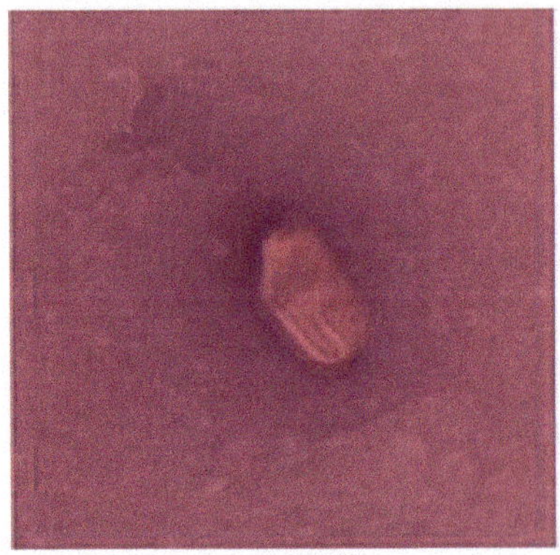

Teratoid

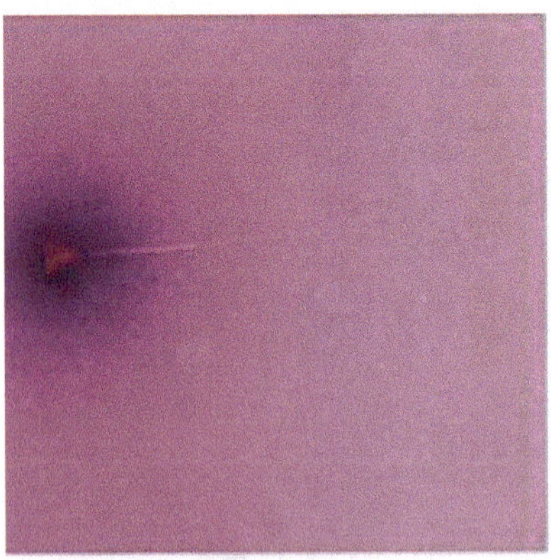

Detached acrosome

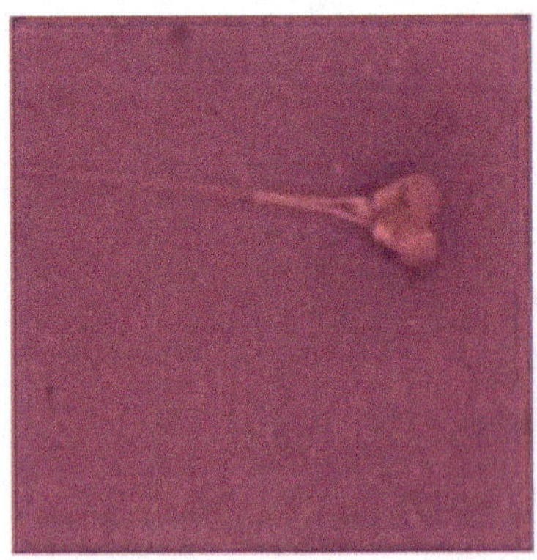

Double head

Abnormalities of bull sperm head

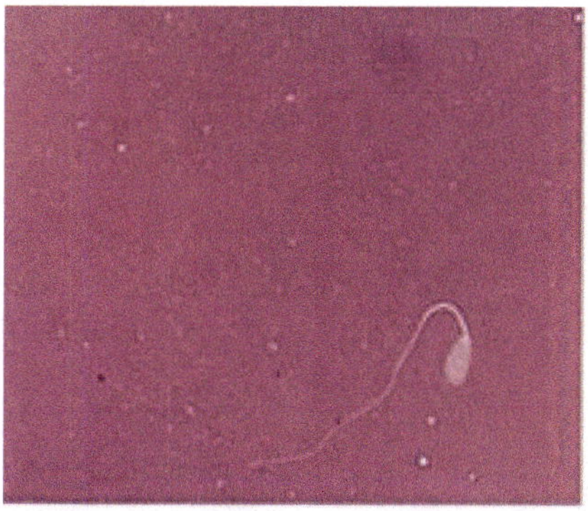

Slightly bent midpiece

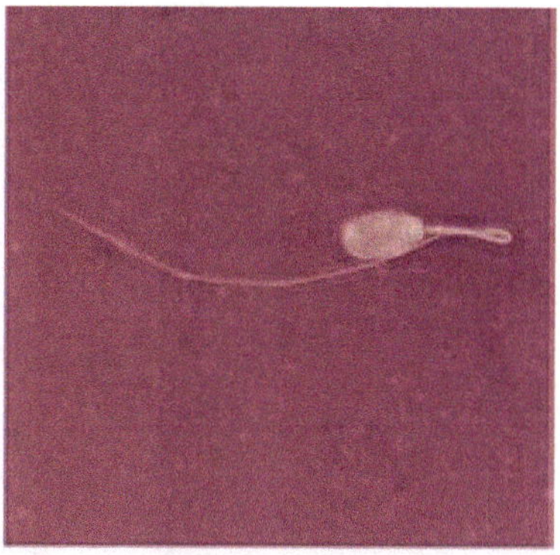

Bent midpiece

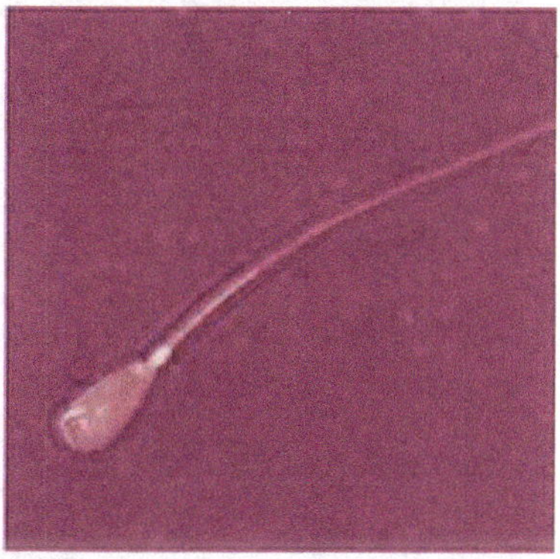

Proximal protoplasmic droplet

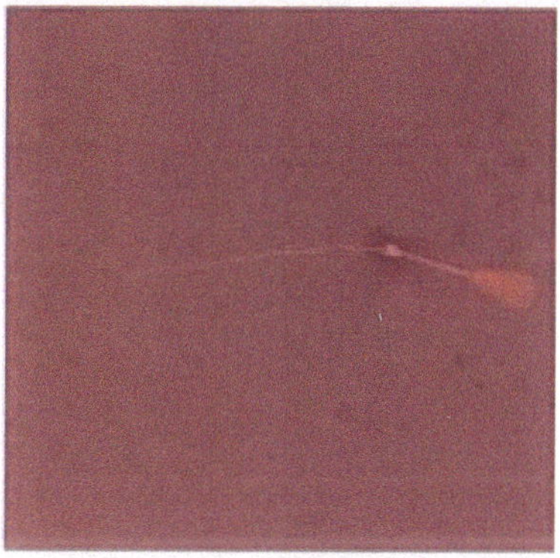

Distal protoplasmic droplet

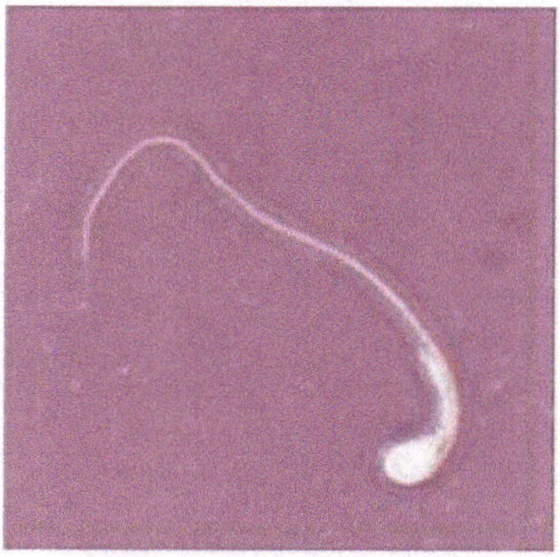

Midpiece bent around
Proximal protoplasmic droplet

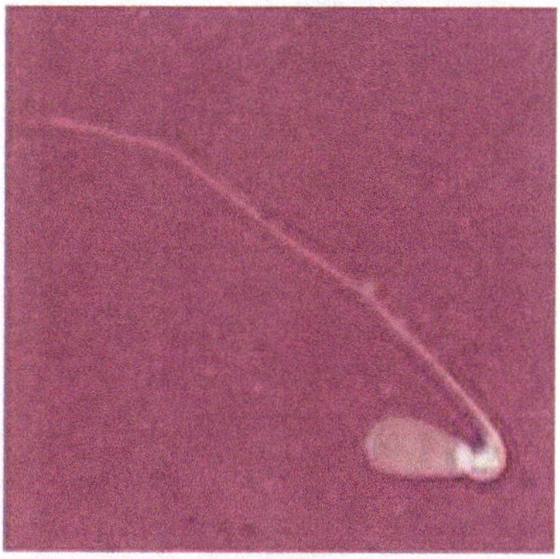

Proximal protoplasmic droplet
with bent midpiece

Abnormalities of bull sperm midpiece

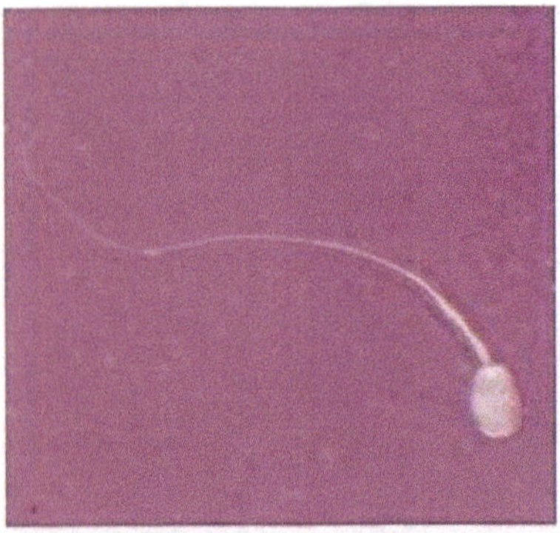

Thickened midpiece

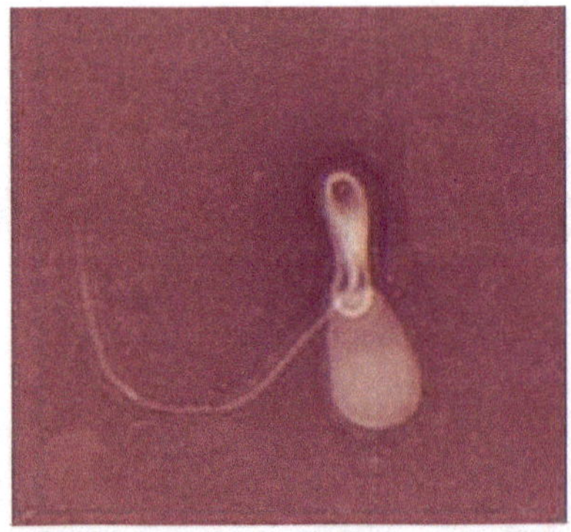

Highly coiled midpiece & tail

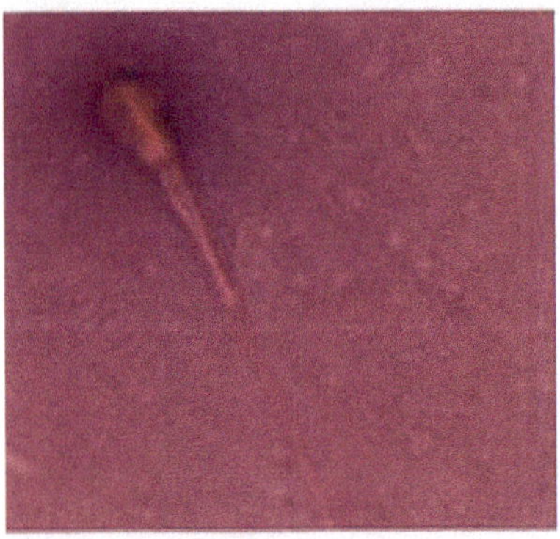

Double midpiece

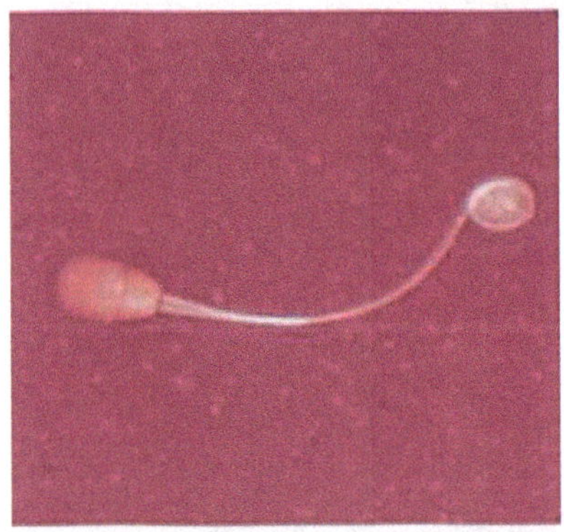

Double midpiece with coiled tail

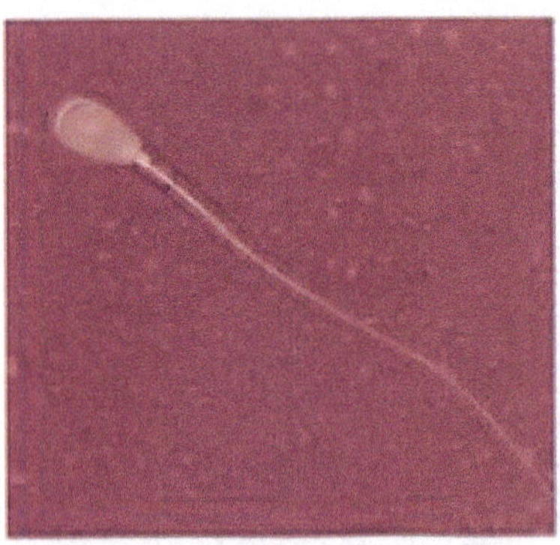

Abaxial implantation

Segmental aplasia

Abnormalities of bull sperm midpiece

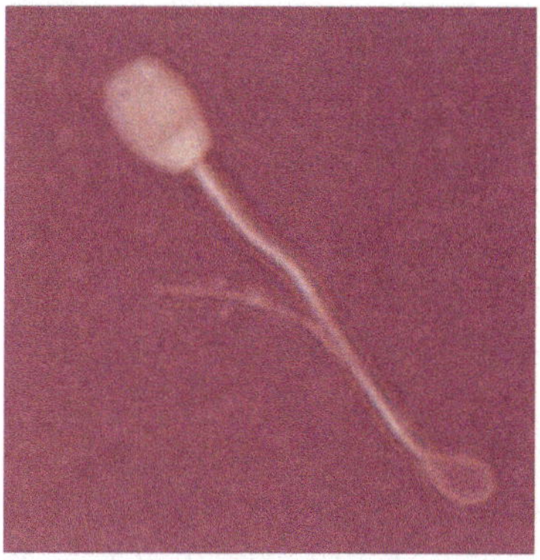

Bent tail

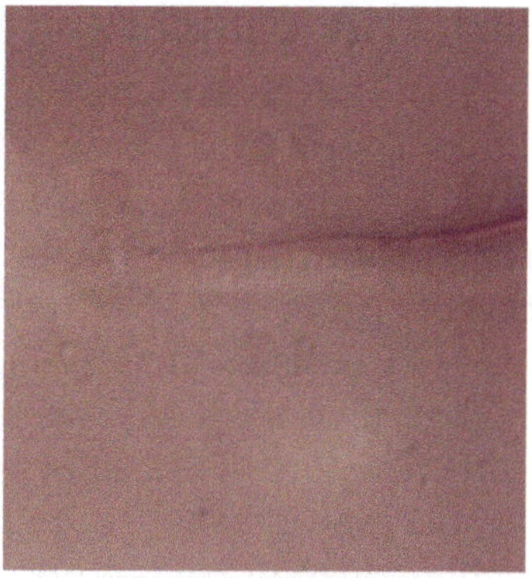

Free tail

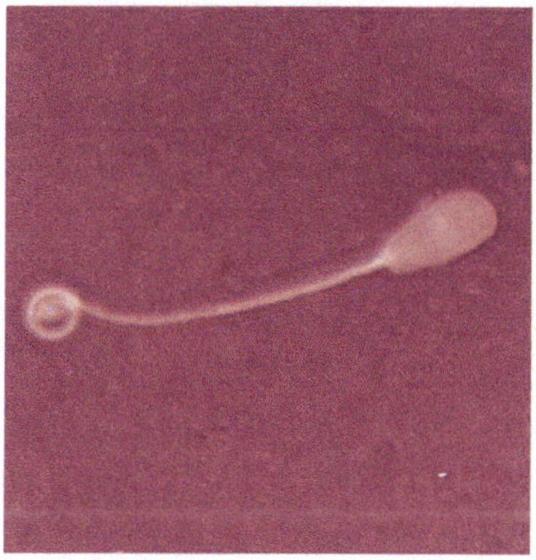

Terminally coiled tail

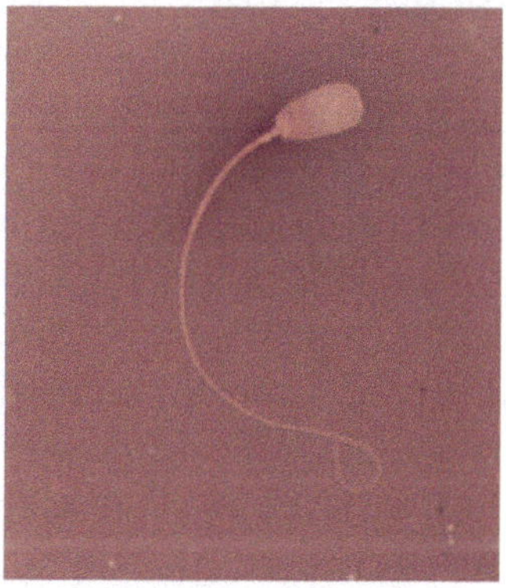

Slightly bent tail

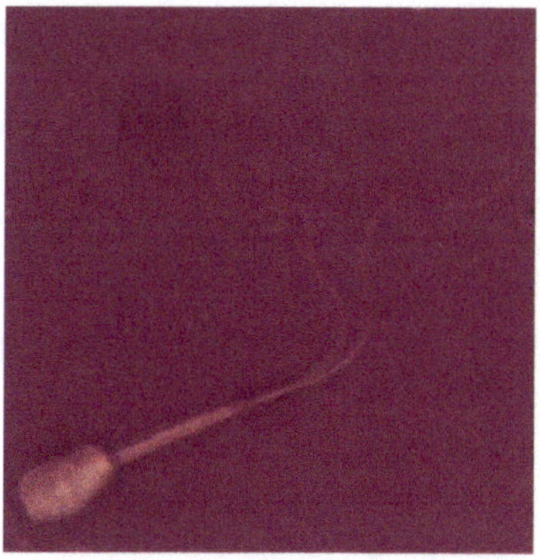

Double tail

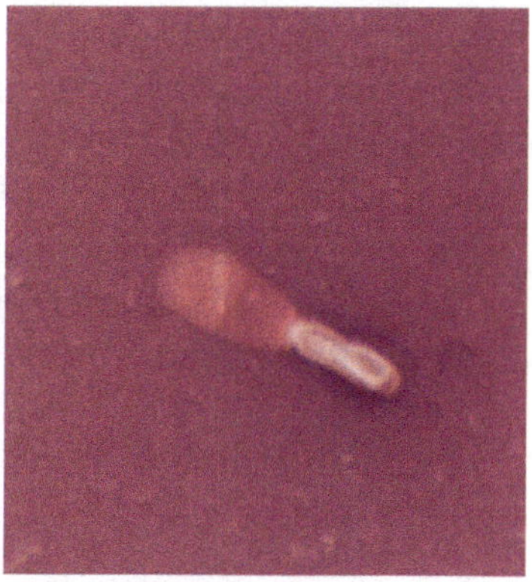

Dag defect

Abnormalities of bull sperm tail

Abaxial Attachment and Multiple Tails

The abaxial attachment is characterized by an attachment of the tail to one side of the base of the nucleus. Usually it is associated with one and rarely two accessory tails i.e. double or triple tails with two or three implantation fossas. However, only abaxial tails without accessory tails are also seen. Little information is available on the fertility of bulls with a high proportion of spermatozoa having abaxial attachment of the tail. The significance of abaxial tail implantation in an affecting fertility has not been fully established. It has been observed that spermatozoa with abaxial tails have normal fertilizing ability, which leads to normal embryonic development, but spermatozoa with accessory tails impair either fertilization or embryonic development.

Accessory Tail

The presence of accessory tail is usually associated with the abaxial attachment of the sperm tail and is always characterized by the presence of a double implantation fossa with a vestige or remnant of a second midpiece or a complete double tail. Rarely, spermatozoa with three implantation fossas and triple tails may also be encountered. Remnants of accessory tails are located lateral to the main tail. Feulgen's staining is superior to eosin-nigrosin staining for revealing accessory tails and double implantation fossas and bull spermatozoa affected with the accessory tail defect (17-22 percent incidence) had low non-return rates.

Other Midpiece Defects

Routine examinations of eosin-nigrosin stained smears of bull semen occasionally reveal very small gaps along the mitochondrial sheath without any impairment of motility and fertility. At times, fracture at these gaps lead to mid piece stumps of varying lengths. This segmental aplasia of mitochondrial sheath commonly leads to rupture of axial fiber bundles at that spot. This may cause loss of principal piece or a sharp bending of the tail. Other midpiece defects such as Coiled, thick, double midpiece etc. are also occasionally seen.

13.2.3 Abnormalities of the Principal Piece or Terminal Piece

Abnormalities of the principal piece can be classified as

(i) Simple bends,
(ii) Coiling and
(iii) Gaps in the fibrous sheath.

Cold shock while semen collection, use of cold diluent, slides, stain etc. can induce bending and coiling of the principal piece. The coiled principal piece defect is not commonly encountered. It is characterized by a tight coiling of the principal piece at various levels distal to annulus affecting the motility of sperm.

Cytoplasmic Droplets

Spherical cytoplasmic masses of 2-3m in diameter are commonly found surrounding the midpiece of a few spermatozoa either at the neck region (proximal protoplasmic droplet) or just proximal to the annulus (distal protoplasmic droplet).

A high proportion of spermatozoa with proximal cytoplasmic droplets in young bulls, indicate premature testicular development, whereas in old bulls it indicates testicular degeneration. Even 5-10 percent of persisting proximal droplets appear to be associated with reduced motility and fertility.

Underdeveloped Spermatozoa

The underdeveloped spermatozoa have several structural aberrations and hence have been named as "teratiod spermatozoa". Occasionally, they may be found in the semen of normal bulls (<1 percent), but their numbers may increase during abnormal spermatogenesis. These sperm almost always take up the eosin stain. They have been classified as major sperm defect.

13.3 Methods for evaluation of sperm morphology

13.3.1 Examination of Wet mounts/ Unstained preparations

13.3.2 Examination of Stained slides

13.3.1 Examination of wet mounts/ unstained preparations

Materials Required

1. Water bath
2. Microscope with 100 X phase contrast/ differential interference contrast objective
3. Microslides
4. Coverslips
5. Scissors

Reagents/ Chemicals Required

1. Buffered formol saline
2. Karnovsky's Fixative

Karnovsky's Fixative

i) Add 0.53 g Paraformaldehyde to 49.4 ml deionized water heated to 60-70°C.

ii) Add 2 drops of 1 N NaOH to clear solution.

iii) Cool to room temperature and add 5.6 ml of 25 % solution of Gluteraldehyde (Sigma) under a fume hood or wear a nasal mask to avoid breathing toxic aldehyde fumes.

iv) Adjust the pH to 7.2 by adding either 0.1 N NaOH or HCl.

v) Mix fixative with 0.2 M Phosphate buffer in the ratio of 1:1.

vi) Check osmolarity.

vii) Store at 5°C.

Buffered Formol Saline

Stock Buffer

Na2HPO4.2H2O (21.682 g/ 500 ml distilled water)	200.00 ml
KH2PO4H2O (22.254 g/ 500 ml distilled water)	80.00 ml
Sodium Chloride Solution	
NaCl	9.01 g
Distilled water up to	500 ml

Buffered Formol Saline

Stock Buffer	100.00 ml
Sodium Chloride solution	150.00 ml
Commercial Formaline	62.5 ml
Distilled water up to	500.00 ml

Preparation of Semen Samples for Examination

Method 1

1. Add 0.1 ml neat semen to 1ml of buffered formol saline or karnovsky' fixative maintained at 30-34 oC (same temperature at which the semen is after thawing/ at the time of evaluation).

2. Mix gently. Semen so preserved can be kept in refrigerator for about a year (for evaluation at a later date) in screw capped vials.
3. Place a small drop (2-4 mm) of preserved semen on a clean microslide.
4. Cover the drop with a coverslip in such a way that the drop spread evenly under the coverslip.
5. Allow about 1 minute time for the drop to spread evenly.
6. Slight pressure may be applied with finger tip to facilitate the drop to spread without injuring the sperm cells.

Method 2

1. Dilute neat semen (1:10) in 2.9 % Sodium citrate.
2. Place about 2-4 mm drop of diluted or frozen-thawed semen on a clean microslide.
3. Place 2-4 mm drop of 0.2 % buffered Gluteraldehyde (in Phosphate Buffer Saline, pH 7.2)
4. Mix the two drops.
5. Cover it with a coverslip.

Observation

Observe at 1000 X magnification under phase contrast microscope or differential interference contrast (DIC). Atleast 200 cells should be counted and classified based on sperm morphology in either one of the following classifications:

1. Primary and Secondary Abnormalities
2. Major and minor sperm defects
3. Head, midpiece and tail defects
4. Compensable and non compensable sperm defects

13.3.2 Examination of Stained Slides

Dilute the neat semen (1:10) in 2.9 % sodium citrate buffer or Tris buffer for assessment of sperm morphology. In case of frozen semen, the straw should be thawed at 37 °C for 30 seconds. Thereafter, a thin smear is prepared in clean microslide prewarmed at 37 °C. Care should be taken to avoid injuries to spermatozoa during smear preparation.

The following staining procedure are used for sperm morphology evaluation:

A. Eosin-nigrosin stain

B. William's Staining/ Carbol-Fuchsin-Eosin Stain

C. Rose Bengal staining

But none of the above mentioned staining procedure can be used for study of acrosomal integrity. The procedure for the study of acrosomal integrity is discussed in the next section.

A. Eosin-Nigrosin Stain

Any one of the following Eosin-nigrosin preparations can be used:

Hancock's (1951) Stain

Eosin Y	5.00 g
Nigrosin	30.00 g
Distilled water	300.00 ml

Dissolve nigrosin in distilled water while stirring and heating until it is dissolved. Later dissolve eosin in the nigrosin solution in a similar manner. Do not boil.

Campbell et al (1960) Stain

Nigrosin Solution	150.00 ml
Eosin Y (GT Gurr)	5.00 g
Stock Buffer Solution	30.00 ml
Stock Glucose Solution	30.00 ml
Distilled Water up to	300.00 ml

Eosin- Aniline Blue Stain **(Shaffer and Almquist, 1948)**

Eosin B	1.00 g
Aniline Blue	4.00 g
M/8 Phosphate buffer	100.00 ml

Stains are dissolved in buffer after heating it to 85°C. Do not boil.

Blom's Stain
Stock Stain
Solution A:

Eosin B	5.00 g
Distilled Water	100.00 ml
Solution B:	
Nigrosin Practical	10.00 g
Distilled Water	100.00 ml
Staining solution	
Solution A	1 part
Solution B	4 part
(pH 9.4-9.9)	

The stain can be kept at room temperature for 2-3 years, without fading and sedimentation.

Barth and Oko's Modifications of Hancock's and Blom's Stain

Eosin Y (color index 45380)	3.30 g
Nigrosin (color index 50420)	20.00 g
Sodium Citrate	1.50 g
Distilled Water	300.00 ml

(Adjust pH to 6.8-7.0; allow the preparation to stand for few days and filter)

- Hancock's and Blom's stain are hypotonic, sodium citrate is added to increase the osmolarity in order to prevent the formation of simple bent tail as artifacts.
- Concentration of the Hancock's and Blom's stain is reduced here to facilitate the stain to dissolve easily.

Smear Preparation and Staining

1. Thaw a straw of frozen semen at 37°C for 30 seconds in a water bath.
2. Maintain a small aliquot of the eosin nigrosin stain at the same temperature for at least 5 minutes before staining.
3. Place 1-2 drops of stain and a very small drop of neat semen (approx. 1/5th of the stain) at one end of a clean microslide. A comparatively larger drop is taken for frozen thawed semen.

4. The drop of stain and semen is mixed with the help of the edge of another microslide without crushing or injuring the sperm cells. French medium straws can also be used for this purpose.
5. Allow this mixture to stand for 1-2 minutes.
6. Draw a smear of this mixture with the help of another slide with smooth edge. The second slide used for preparing the smear should be drawn smoothly by holding it at an angle of 45° and ensuring that the sperm stain mixture is in the inner angle of the two slides. Glass slide with rough edges should not be used for drawing smears as it may lead to detached heads, free tails, broken neck and damaged acrosomes.
7. Dry the smear quickly by blowing hot air with blower or hair dryer or by keeping at warm stage maintained at 37°C. In no case should the smear be dried over a flame.

B. William's staining/ Carbol-Fuchsin-Eosin Stain

Reagents/ Chemicals Required:

Loeffler's Methylene Blue

Loeffler's Methylene Blue

(A) Stock Methylene Blue Solution

Methylene blue	10.00 g
Alcohol 96 %	100.00 ml

(B) Staining Solution

Stock Methylene Blue Solution	30.00 ml
Potassium Hydroxide (0.01 %)	100.00 ml

Dilute it to 1:3 (V/V) in glass distilled water immediately prior to staining.

Note : After using for 2 weeks, Williams stain in the coupline jar should be filtered through filter paper grade – 1 to use it again. Otherwise, some sperms which detach from smear may get stuck to other smears (these smears will look much more red) and will interfere with morphological study.

Procedure for Smear Preparation and Staining

1. Prepare smear of neat semen on a grease free slide. Dry in air
2. Fix in flame.
3. Dip in absolute alcohol for 3-4 minutes.
4. Put over a filter paper for drying.
5. Dip in 5 % Chloramine-T solution for 2-5 minutes until the mucous is removed.
6. Rinse with Distilled water and then give a quick dip in 96 % alcohol.
7. Stain with Carbol-Fuchsin-Eosin Stain for 10 minutes.
8. Wash in tap water for 30 seconds. Some workers prefer to counter stain the slide stained as above with Loeffler's methylene Blue for 1-5 seconds. This step is optional and can be avoided if sufficiently good staining with Carbol-Fuchsin-Eosin Stain is attained.
9. Dry in room temperature for about half an hour.

C. Rose Bengal Staining

Reagents/ Chemicals Required

3 % Rose Bengal Stain	
Rose Bengal	3.00 g
Commercial Formaline	1.00 ml
Distilled Water up to	100.00 ml

Filter through filter paper grade 1 and store in sterilized bottle

Procedure for Smear Preparation and Staining

1. Place a small drop of semen on a clean grease free microslide.
2. Prepare semen smear with the help of another slide having smooth edges
3. Dry the smear in air
4. Stain in 3 % Rose Bengal stain for 5-6 minutes.
5. Wash the excess stain with distilled water.
6. Dry in air.

Observation

Observe at 1000 X magnification under oil immersion objective of a microscope. At least 200 cells should be counted and classified based on sperm morphology in either one of the following classifications:

1. Primary and Secondary Abnormalities
2. Major and minor sperm defects
3. Head, midpiece and tail defects
4. Compensable and non compensable sperm defects

Calculate per cent abnormal spermatozoa as per the following equations:

$$\text{Per cent abnormal sperms} = \frac{\text{Total nos. of abnormal sperms counted}}{\text{Total nos. of sperm cells counted}} \times 100$$

Inference

American Society for Theriogenology considers semen samples containing more than 70 % normal sperm cell to be fit for AI. Bull producing semen with more than 70 % normal sperm irrespective of the type of abnormality are considered satisfactory breeders.

Classification of Sperm Abnormalities

1. Primary and Secondary Sperm Abnormalities

Primary Sperm Abnormalities

- Underdeveloped
- Double Forms
- Acrosome Defect (e.g., knobbed acrosome)
- Narrow heads
- Crater/diadem defect
- Pear-shaped defect
- Abnormal contour
- Small abnormal heads
- Free abnormal heads
- Abnormal midpiece
- Proximal droplet

- Strongly folded or coiled tail
- Accessory tails

Secondary Sperm Abnormalities

- Small normal heads
- Giant and short broad heads
- Free normal heads
- Detached, folded, loose acrosomal membranes
- Abaxial implantation
- Distal droplet
- Simple bent tail
- Terminally coiled tail

Other Cells

- Epithelial cells
- Erythrocytes
- Medusa formations
- Sperm precursor cells
- Round cells
- White blood cells

2. Major and Minor Sperm Defect (Blom, 1972, 1977a)

S. No.	Major Sperm defects	Minor Sperm defects
a.	Underdeveloped Spermatozoa	Narrow head
b.	Double sperm forms	Small normal head
c.	Knobbed Acrosome	Giant and short broad head
d.	Active tails/ Decapitated sperm	Free normal head
e.	Diadem defect	Detached Acrosomal membranes
f.	Pear shaped head	Abaxial implantation of tail
g.	Narrow at base	Distal droplet
h.	Abnormal contour	Simple bent or coiled tail
i.	Small abnormal head	Terminally coiled tail
j.	Free pathological head	
k.	Cork screw defect	
l.	Stump tail	
m.	Proximal droplet	
n.	Pseudo droplet	
O.	Dag defect	

3. Abnormalities of Sperm Head, Midpiece and Tail

S No.	Head Abnormalities	Midpiece Abnormalities	Tail Abnormalities
a.	Free/ detached heads	Bent mid piece	Bent tail
b.	Knobbed Sperm	Cork screw defect	Terminally coiled tail
c.	Detached acrosome	Pseudodroplet defect	Distal droplet
d.	Diadem defect	Abaxial implantation of midpiece	Dag defect
e.	Pyriform head	Proximal droplet	Stump tail defect
f.	Microcephalic sperm	Double midpiece	Double tail
g.	Mega cephalic sperm	Bowed midpiece	Distal droplets
h.	Abnormal contour	Segmental aplasia of mitochondrial sheath	
i.	Double head	Swollen midpiece	
j.		Filliform neck	

Note- One or more than one defect may appear in association in single spermatozoa.

Chapter 14

In vitro Fertility Tests

Only few single sperm viability parameters show a significant correlation with fertility of semen samples. Therefore, need to explore functional in vitro fertility tests having capability to disclose the ability of spermatozoa to undergo complicated processes such as capacitation, binding of zona pellucida, acrosome reaction, fertilization of Oocyte and induction of embryo development invitro have been designed and explored for their correlation with fertility achieved after artificial insemination.

In vitro fertility tests used in evaluation of semen includes:

14.1 Estimation of membrane integrity

- 14.1.1 Morphological integrity
- 14.1.2 Functional integrity
- 14.1.2.1 Hypoosmotic swelling test
- 14.1.2.2 Acrosome integrity

14.2 Cervical Mucus Penetration Test

14.3 Zona Binding Assay

14.4 Heterospermic insemination and competitive fertilization

14.5 Hemizona assay

14.6 Chromatin Analysis

14.7 Oviductal epithelial cell explant test

14.1 Estimation of Membrane Integrity

Evaluation of membrane integrity of the spermatozoa can be done by the following ways:

14.1.1 Morphological integrity

14.1.2 Functional integrity

14.1.2.1 Hypoosmotic swelling test

14.1.2.2 Acrosome integrity

14.1.1 Morphological Integrity / Viability.

Plasma membrane integrity is essential for the maintenance of sperm viability. Several staining methods have been developed to detect disruption in the plasma membrane. The principle of these techniques is dye exclusion, which means that these stains can not penetrate through the intact membrane, but they color the spermatozoa if the plasma membrane is damaged, due to the affinity of the stains to bind to the nucleus. First, eosin-nigrosin staining was used to differentiate "live" and "dead" mammalian spermatozoa. This technique is easy to use, no special microscope is required and it gives reliable results. Since fluorescent microscopy and flow cytometry become more frequently used in semen evaluation, new fluorescence supravital staining techniques have been developed. The most commonly used fluorescence staining involves the use of Hoechst 33258 (H258).

Hoechst 33258 (H258) dual DNA stain consisting of propidium iodide (PI) combined with carboxyfluorescein diacetate (CFDA) or with SYBR-14, or carboxydimethylfluorescein diacetate (CDMFDA). Propidium iodide is a DNA-specific stain, which enters the dead or moribund sperm cells. CFDA and CDMFDA are nonspecific esterase substrates, which are non fluorescent and can penetrate through the membranes. Entering the sperm cell, nonspecific esterases readily hydrolyse them, resulting in a highly fluorescent, membrane impermeable green fluorescent stain. Dual staining techniques classify the sperm population into three categories:

(1) live spermatozoa emitting green fluorescence resulting from positive staining by CFDA or SYBR-14,

(2) dead spermatozoa with nuclei emitting red fluorescence resulting from positive PI and negative CFDA or SYBR-14 staining,

(3) moribund spermatozoa with nuclei emitting green and red fluorescence resulting from positive staining by PI and CFDA or SYBR-14 in the same time.

14.1.2 Funcional Integrity

14.1.2.1 Hypo Osmotic Swelling Test (Host)

The routine methods of semen evaluation have limited accuracy in predicting the fertilizing potential of spermatozoa in the semen sample, hence in vitro fertility tests can prove to be useful in assessment of fertilizing potential of sires and can act as an effective supplement to the routine semen evaluation methods in accurate fertility evaluation of semen. Hypo osmotic swelling test is a relatively simple test to evaluate the functional integrity of the spermatozoal membrane. Biochemically active sperm membrane is required for the process of capacitation, acrosome reaction and the binding of the sperm cell to the oocyte surface. Jeyendran developed this assay for human spermatozoa in 1984 and recently HOS has been used in bulls, in boars, in dogs and in stallion. First, Drevius and Eriksson observed that spermatozoa exposed to hypoosmotic conditions undergo morphological alteration and their size is increased. During the HOS test, the biochemically active spermatozoa, due to the influx of water, will undergo swelling and increase in volume to establish equilibrium between the fluid compartment within the spermatozoa and the extra cellular environment. This volume increase is associated with the spherical expansion of the cell membrane covering the tail, thus forcing the flagellum to coil inside the membrane. Coiling of the tail begins at the distal end of the tail and proceeds towards the mid-piece and head as the osmotic pressure of the suspending media is decreased. The plasma membrane surrounding the tail fibers appears to be more loosely attached than the membrane surrounding the head.

Hypoosmotic swelling test is based on the principle that if sperm cell with functional and intact plasma membrane is exposed to hypoosmotic environment, it will response by swelling/ coiling of tail. It can also be termed as membrane stress test. This is being nowadays routinely used to test in vitro fertility of spermatozoa. The semipermiable sprematozoal membrane allows selective passage of water through it along an osmotic pressure gradient. If the sperm cells are placed in a hypoosmotic solution, water passes into the spermatozoa to reestablish the equilibrium between the extra cellular and intracellular fluid compartments resulting in the ballooning of the sperm head and deformation of the tail. The hypoosmotic swelling test relies on this deformation of the tail, apparent in spermatozoa that were intact when placed in solution of osmolarity less than 290 mOsm kg-1. The deformation of the tail is manifested in the form of bending and coiling and was evident in spermatozoa with intact membranes. This is due to increase in the cell volume of the spermatozoa and the expansion of the membrane covering the sperm tail causing the flagellum to coil inside it. Transition electron microscopy revealed that there is coiling of the flagellum inside the plasma membrane and the flexible motor apparatus of the tail is forced by the cell membrane to bend and coil. The coiling of the flagellum in response to lowered osmotic pressure has been reported to be positively correlated with fertilizing ability of the semen.

The optimal hypo-osmotic medium should exert an osmotic stress large enough to cause an observable increase in volume, but small enough to prevent the lysis of the sperm membrane. Hypo-osmotic mediums contain mainly fructose and sodium citrate mixed in equal proportions, or fructose, sodium citrate, lactose and sucrose alone. Optimal results can be obtained in the range of 50 to 150 mOsm/L osmolality.

A significant correlation of the HOS test exist with the sperm motility and the per cent ovum penetration, sperm morphology and per cent live sperm. Reports of correlation of this test with denuded hamster oocyte penetration test are also available.

This test has been designed for diverse mammalian species such as cattle, horses, pigs and canines. The modified hypoosmotic stress test is widely used in the evaluation of bull semen, which is closely correlated with fertilization and non return rate.

Procedure of Hypo Osmotic Swelling Test

The Hypo osmotic swelling can be performed by either of the following methods :

A. With Hypo osmotic solution

B. With distilled water

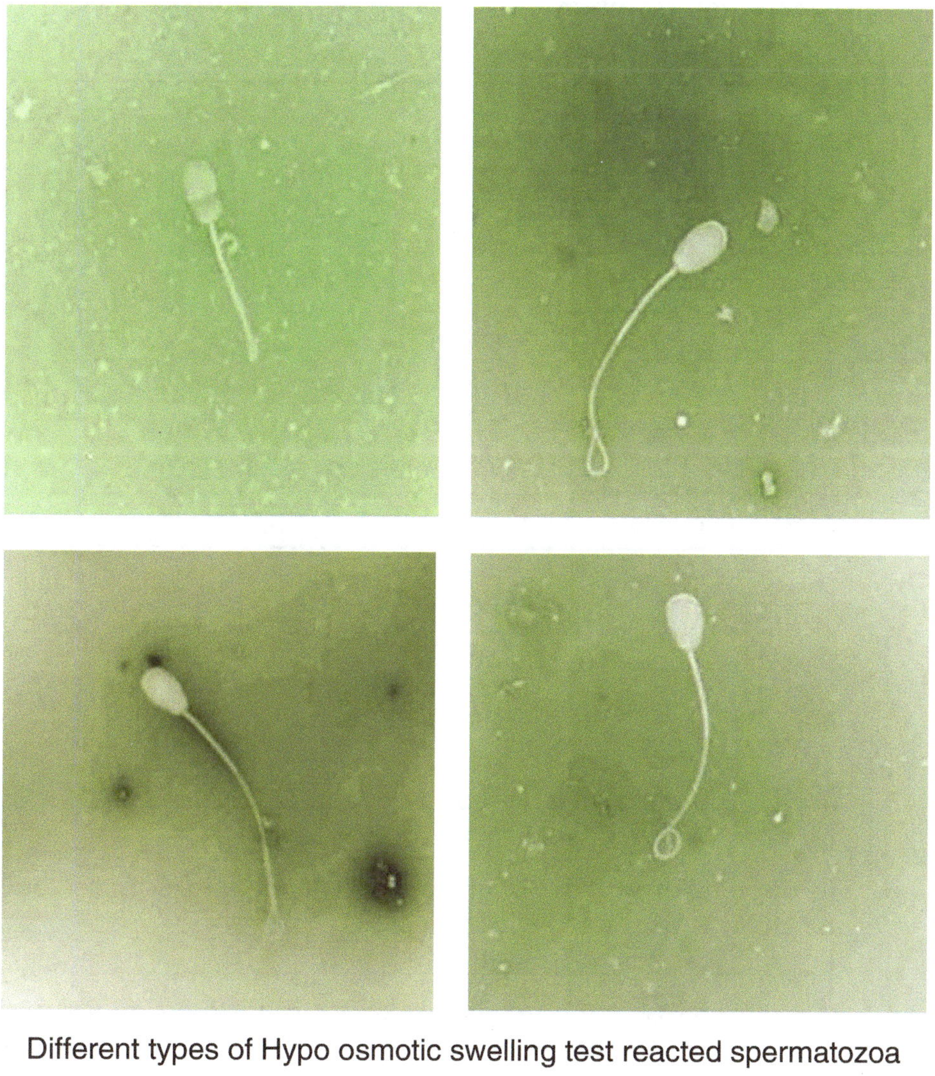

Different types of Hypo osmotic swelling test reacted spermatozoa

A. With Hypo Osmotic Solution

Materials Required

1. Phase contrast microscope
2. Water bath
3. Sugar tube/ Screw capped vial
4. Microslides
5. Coverslip

Chemicals/ Reagents Required

1. Hypo osmotic solution
2. Buffered formol saline

Hypo Osmotic Solution

	100 mOsm	150 mOsm
D-fructose	0.9 g	1.351
Tri sodium citrate	0.490 g	0.735 g
Distilled water up to	100 ml	100 ml

Buffered Formol Saline

Na2HPO4.2H2O	6.194 g
KH2PO4H2O	2.5432 g
NaCl	5.406 g
Commercial Formaline	125.00 ml
Distilled water up to	1000.00 ml

Procedure

1. Add 0.1 ml of semen to 1 ml of hypo osmotic solution (100 mOsm) in a sugar tube. Mix gently.
2. Incubate the semen and hypo osmotic solution mixture at 37°C for 30 minutes.
3. After incubation mix the contents gently and add 0.1 ml of buffered formal saline or formaline. Mix gently
4. Place a small drop of the incubate on a clean greese free microslide and cover it with a cover slip.
5. The samples can be examined at 400 X magnification of a phase contrast microscope immediately or kept at room temperature for examination later on.

B. HOST with Distilled Water

Materials Required

1. Phase contrast microscope
2. Water bath
3. Sugar tube/ Screw capped vial
4. Microslides
5. Coverslip

Chemicals/ Reagents Required

Buffered Formol Saline

1. Add 0.1 ml of semen to 0.4 ml of distilled water in a sugar tube. Mix gently.
2. Incubate the semen and hypo osmotic solution mixture at 37 oC for 15 minutes.
3. After incubation mix the contents gently and add 0.1 ml of buffered formal saline or formaline. Mix gently
4. Place a small drop of the incubate on a clean greese free microslide and cover it with a coverslip.
5. The samples can be examined at 400 X magnification of a phase contrast microscope immediately or kept at room temperature for examination later on.

Observation

The evaluation should be done at 400 X magnification of a phase contrast microscope by counting at least 200 spermatozoa. The spermatozoa with swollen area at the tip of the tail, shortened and thickened tail should be counted as HOS reacted whereas others are counted as non reactors and per cent HOS reacted spermatozoa is Calculated.

Inference

Frozen semen samples with HOS reacted sperm less than 50 % should not be used for AI

Staining HOST Slides with Rose Bengal

The slides prepared to evaluate HOS per cent may be stained for better clarity, photography or examination at a later date.

Procedure

Either of the above two procedure may be adopted till incubation stage. Following incubation a thin smear of the incubated mixture is prepared on a clean greese free microslide. Stain the smear with 3 % Rose Bengal stain. Wash the slides with distilled water to remove the excess stain. Examine at 1000 X magnification of a microscope under oil immersion.

14.1.2.2 Acrosome Integrity

Acrosome is a cap like structure on the head of the spermatozoa. It covers approximately 60-70 % of bovine sperm head. Acrosome contains enzymes which are instrumental in penetration of the ovum. Therefore, acrosome integrity needs to be maintained to achieve normal fertility. Significant changes in the plasma membrane and outer acrosomal membrane are observed during capacitation. These changes constitutes the process of acrosome reaction. The changes during the acrosome reaction include fusion at multiple points between the two membranes and formation of vesicles made up of fragments of the two membranes. The sperm must be able to undergo these changes in the female reproductive tract to attain optimum fertilization potential, which requires the acrosome to be structurally intact and biochemically functional. Various physical and chemical factors influence the acrosome integrity. Acrosome can also be damaged during the process of freezing and thawing. Hence, evaluation of acrosomal integrity should be integral part of semen quality testing. To fertilize an oocyte, spermatozoa must possess an intact acrosome. Shortly before fertilization, sperm cell undergoes acrosome reaction which is required for penetration through the zona pellucida and for fusion with the oocyte plasma membrane. Several techniques have been used to differentiate the acrosome intact from acrosome-reacted spermatozoa. The physiological acrosome reaction is a well coordinated process that only occurs in living spermatozoa and is called"true acrosome reaction". In contrast, the loss of acrosome may also occur due to degenerative changes in the membrane (freezing), which is called "false acrosome reaction". To differentiate these two reactions, acrosomal stainings are combined with supravital stains. Acrosomal staining alone can be used for:

(1) Evaluation of semen for acrosomal abnormalities,
(2) Evaluation of the effect of cryopreservation on the acrosome,
(3) Evaluation of acrosomal status during in vitro capacitation.

The acrosomal status of sperm from species with large acrosome (e.g. bull, hamster) can be assessed by phase contrast or differential interference microscope without staining. In other species, including mouse, human, and stallion, the unstained acrosome is too small to be visualized accurately. For this reason, specific acrosomal labeling techniques have been used to enhance the visualization of the acrosome with bright-field or fluorescence microscopy. For bright-field microscopy, the triple stain technique (trypan blue, Bismark brown and rose bengal) is used most frequently. For fluorescence microscopy, there are two kinds of fluorescent probes. The first type detects intracellular acrosome-associated materials, and requires cell permeabilization before labeling. This category consists of lectins and antibodies to intracellular antigens. The second type of fluorescence probe consists of reagents that are used without permeabilization of cells, such as chlortetracycline (CTC) and antibodies to externally exposed antigens. Lectins bind to the glycoconjugates of the acrosomal matrix or outer acrosomal membrane. First, a highly toxic lectin (Ricinus communis agglutinin-II), labeled with fluorescein isothiocyanate (FITC), was used, but later the technique was modified and Pisum sativum agglutinin (PSA) became the most wildly used lectin since then. A number of staining methods to assess acrosomal changes in stallions have been described. These include the use of chlortetracycline, monoclonal antibodies, FITC-PSA and fluorescein isothiocyanate conjugated Arachis hypogea agglutinin (FITC-PNA). Spermatozoa stained with FITC-PSA have different staining patterns, depending on the acrosomal status. Bright fluorescence over the acrosomal cap indicates "acrosome intact" sperm. No fluorescence at all or limited fluorescence at the equatorial segment indicates complete loss of acrosomal cap, either due to complete acrosomal reaction, called "acrosome reacted", sperm or due to cell death caused by membrane degeneration. These staining patterns do not give precise information about the process of acrosome reaction. However, the staining patterns of FITC-PNA clearly demonstrate the different stages of acrosome reaction. Appearance of the "acrosome intact" and "acrosome reacted" spermatozoa are the same as with PSA staining, but "acrosome reacting" cells can be differentiated by patchy disrupted fluorescence over the acrosomal cap. FITC-PSA and FITC-PNA staining can also be used without permeabilization of cells. This method allows the evaluation of acrosomal membrane injuries of spermatozoa. While spermatozoa with damaged plasma and acrosomal membranes become fluorescent over the acrosomal cap and the equatorial region, spermatozoa with intact membranes, impermeable to these lectins, do not show fluorescence after exposure.

Evaluation Methods for Acrosome Integrity

A. Giemsa Staining
B. Wet Mounts

A. Giemsa Staining

Materials/ Equipments Required

1. Microscope with 100 X oil immersion objectives
2. Water bath
3. Centrifuge
4. Hot plate maintained at 37 oC
5. Microslides
6. Coupline jars
7. Pastle and Mortar
8. Microcentrifuge tube
9. Sugar tube

Reagents/ Chemicals Required

1. Giemsa Stain
2. Fixative (5 % Formaldehyde or buffered formol saline)

Giemsa Stain (Watson, 1975)

Giemsa Stock Solution

(i) Dissolve 1 g Giemsa in 66 ml Glycerol in Mortal.

(ii) Pour the mixed liquid in the flask and heat it on magnetic stirrer on 60 oC for one hour.

(iii) Cool it at room temperature and add 84 ml of methanol in the flask and mix it.

(iv) Stock solution of 150 ml of Giemsa is ready.

Giemsa Working Solution

It should be prepared just before use in either one of the following ways

a.	Giemsa Stock solution	3 ml
	Sorensen's 0.1 M phosphate buffer, pH 7	2 ml
	Distilled water	35 ml

Contd...

Contd...

b.	Giemsa Stock solution	1 ml
	Distilled water	9 ml

Fixatives

5 % Formaldehyde

Formaldehyde	5.00 ml
Distilled water	95.00 ml

Buffered Formal Saline

Na2HPO4.2H2O	6.194 g
KH2PO4H2O	2.5432 g
NaCl	5.406 g
Commercial Formalin	125.00 ml
Distilled water up to	1000.00 ml

Procedure for Smear Preparation and Staining

1. Thaw a semen straw.
2. Add 2 ml of tris buffer to it and centrifuge at 1500 rpm for 5 minutes. Discard the supernatant and add 2 ml Tris buffer.
3. Take greese free microslide prewarmed at 37 °C. Prepare a thin smear of neat semen diluted 10 times in 2.9 % sodium citrate buffer or frozen thawed semen without further dilution on the microslide.
4. Air dry the smear.
5. Fix the smear by either of the following methods.
   a. In buffered formal saline for 30 minutes
   b. In 5 % formaldehyde at 37°C
   c. By heating over a flame
6. Wash the fixed smear in running tap water if fixed by the first two methods.
7. Air dry the smear.
8. Stain with Geimsa working stain for 3 hours. Rinse in two or three changes of distilled water.
9. Air dry the slide.

Observation

Observe at least 100 cells at 1000 X magnification under under oil immersion objective of a microscope. Preferably 200 or more cells should be examined if higher accuracy is desired. Classify the spermatozoaon the basis of acrosomal integrity as follows:

1. Normal Acrosome
2. Abnormal Acrosome
   a. Ruffled
   b. Incomplete
   c. Knobbed
   d. Completely lost

Wet Mounts

Materials Required

1. Phase contrast Microscope
2. Water bath
3. Hot plate maintained at 37°C
4. Microslides
5. Coverslips
6. Sugar tube

Chemicals/ Reagents Required

1. 0.2 % Gluteraldehyde in PBS
2. 0.15 M Sodium Cocodyl

1. Mix 0.5 ml of frozen thawed semen with 1 ml of 0.2 % Gluteraldehyde in PBS or 0.2 % Gluteraldehyde in three parts of PBS and 1 part 0.15 M Sodium Cocodyl.
2. Place a 2-4 mm size drop of this mixture on a clean grease free microslide and cover it with a coverslip.
3. Allow the drop to spread evenly under the coverslip to get a very thin smear.

Observation

Observe the slide under a phase contrast microscope. To record the following:

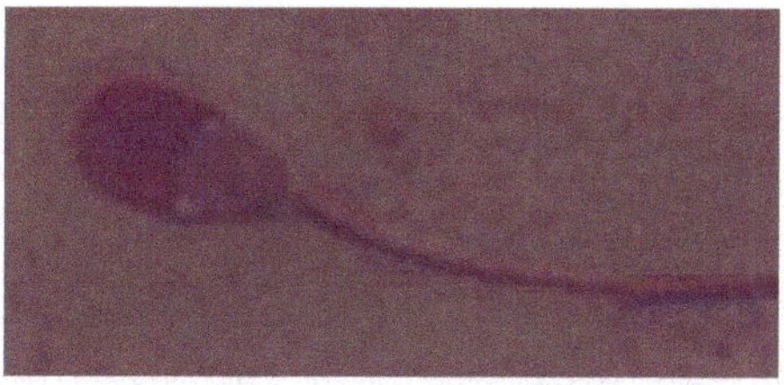

Normal intact acrosome

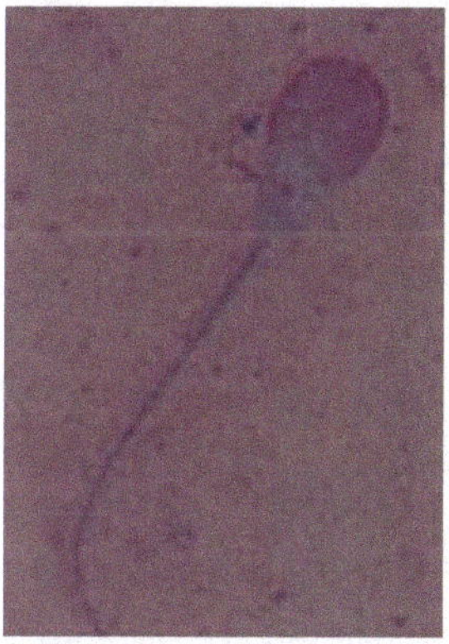

Swollen acrosome

Ruffled acrosome

Loose acrosome

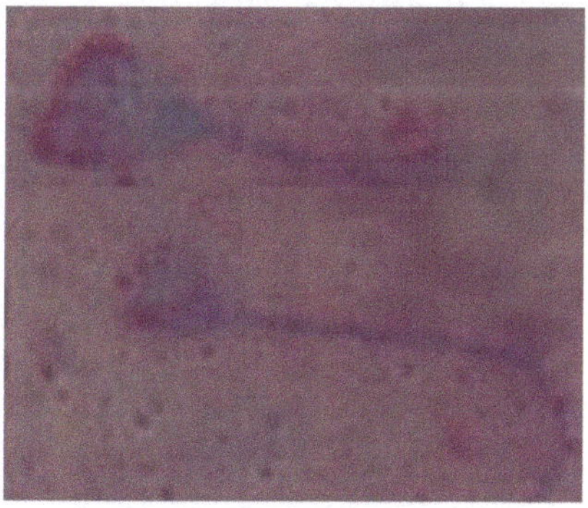

Broken / Detached acrosome

Various types of Acrosomal abnormalities

a. Dark ridge at apex of the sperm cell - Intact Acrosome

b. Dark ridge absent at apex of the sperm cell - Swollen Acrosome

c. Faint ruffled line at the apex of the sperm nucleus - Broken membrane/ Non intact Acrosome

Inference

Frozen semen samples with less than 50 % normal intact acrosome should not be used for AI.

14.2 Cervical Mucus Penetration Test

The viability and migration of the spermatozoa in the female reproductive tract is most essential factor for higher conception. The ability of spermatozoa to penetrate cervical mucus has been used to obtain an idea about the viability of the spermatozoa in several species. The semen deposited by natural service or AI has to be transported to the site of fertilization in the female genital tract. The sperm cell are transported through the cervical mucus, which is biochemically different from the seminal fluid. The cervical mucus penetration test gives a fairly good idea about the penetration capability of the spermatozoa through the cervical mucus. A highly significant positive correlation of ability of the spermatozoa to penetrate the cervical mucus and Acrosome integrity and structural integrity of the spermatozoa has been reported. The limitation of this test lies on the fact that stage of estrum of the animal from which the mucus is collected. The storage time of the mucus also bears significant impact on the test results. Cervical mucus collected from animals which are not in proper stage of estrum and stored for prolonged period before being used in the test may lead to erratic results.

Materials Required

1. Water bath
2. Microscope
3. Cervical mucus of estrus animal
4. PVA powder
5. Non heparinised capillary tube (80 mm X 0.8 mm)
6. Sugar tube
7. Petri dish (90 mm)
8. Syringe - 5 ml
9. Scalp vein set
10. Scissors
11. China Clay

Test Procedure

1. A non heparinised capillary tube of 80 mm length and 0.8 mm diameter is taken.
2. One end of the capillary tube is sealed with PVA powder.
3. Place around 2 ml cervical mucus in a Petri dish. Warm it to room temperature before use.
4. The cervical mucus is drawn in the capillary tube with the help of a scalp vein set attached to the capillary at one end (after removing the needle) and to a 5 ml syringe at the other end. The trailing mucus is then cut with scissors, leaving a small amount protruding from the filled end of the tube.
5. Remove the capillary tube from the loading manifold.
6. The open end of the capillary is then plugged with china clay.
7. The mucus filled capillary tubes are then allowed to stand at 37 °C for 5 minutes.
8. Thaw a frozen semen straw and empty it's contents into a serum vial of 5 ml capacity.
9. Suspend the capillary tube vertically into the serum vial and incubate it at 37 °C for 30 minutes.
10. After 30 minutes of incubation remove the capillary tube from the vial and wipe it with tissue paper to remove extra sperms adhering to the capillary tube.

Observation

Place the tube on a graduated microscope slide and observe the spermatozoa swimming in the mucus under 400 X magnification of a phase contrast microscope.

Measure the distance traveled by the spermatozoa in mm. This distance is called Sperm Penetration Distance (SPD)

Inference

Semen sample with SPD 30 mm or more are considered to have better chance of fertilization.

14.3 Zona Binding Assay

The conventional parameters of semen evaluation have limitations in predicting the fertilization potential of the semen. Therefore, the routine semen evaluation tests needs to be combined with in vitro fertility tests to get a better prediction of sperm fertilizing potential. Spermatozoon's close association with zona pellucida of ova can be used as a vital characteristic exhibiting possible fertility of sperm. This association exhibits species specificity which is termed as attachment. Based on the principle of species-specific binding of sperm with ova, a test is developed to evaluate in vitro fertility.

Materials Required

1. Test tube/ Sugar tube
2. Scissors
3. Centrifuge and centrifuge tubes
4. Micropippette
5. Water bath
6. CO2 Incubator
7. Bovine ovaries
8. Needle and syringe
9. Petridish
10. 35 mm multiwell petridish
11. Narrow bore Pasteur pippette

Reagents/ Chemicals Required

1. mBWW medium
2. BSA
3. Phosphate buffer saline
4. Mineral oil
5. Parraffin/ Vaselin
6. 2-5% gluteraldehyde
7. Hochest 33342 dye

Modified Bigger- Witten- Whittengham (mBWW) Medium (g/ L)

NaCl	5.540
KCl	0.356
CaCl2.2H2O	0.250
KH2PO4	0.162
MgSO4.7H2O	0.294
NaHCO3	1.500
Glucose	1.000
Na-Pyruvate	0.028
Albumin	3.000
HEPES Sodium	1.280
HEPES free acid	1.200
Na Lactate	0.370 ml
Streptopenicillin (50 mg ml-1)	1.000 ml
0.2 % Phenol Red	0.500 ml

Phosphate Buffered Saline (PBS)

Solution A: Sodium Phosphate monobasic anhydrous (NaH2PO4) 0.800 g
Glass Distilled Water up to 100.00 ml

Solution B: Sodium Phosphate dibasic anhydrous (Na2HPO4) 0.947 g
Glass Distilled Water up to 100.00 ml

Phosphate Buffered Saline (PBS)

Solution A	30.00 ml
Solution B	70.00 ml

Test Procedure

A. Sperm Preparation

1. Two ml of frozen semen is taken and is diluted in 4 ml of mBWW medium.
2. The diluted semen is centrifuged for 15 min. at 300 g.
3. The supernatant is discarded and 2 ml of medium is added very slowly so that the sperm pellet is not disturbed.
4. Incubate at 37°C in air for one hour in slanting position for the motile spermatozoa to swim up.
5. The supernatant is collected and centrifuged.
6. Discard the supernatant.
7. The sperm pellet is resuspended in mBWW (containing 6 ml ml-1 BSA) to a final concentration of 10^6 spermatozoa per ml.

B. Preparation of Oocytes

1. Bovine ovaries are obtained from a slaughter house one day before the experiments and are kept in phosphate buffered saline (PBS) at 5°C.
2. Antral follicles with a diameter of 4 to 8 mm are punctured and the aspirates of the follicular fluid are pooled in a petridish.
3. Recover the cumulus oocytes complexes from the follicular fluid, collected in PBS and keep it for four minutes to remove cumulus cells. Mature oocytes have more cumulus mass.
4. The denuded oocytes are washed with PBS and kept at room temperature until they are incubated with spermatozoa.

C. Zona Binding Assay

1. Ten microliter droplet of 10^6 motile spermatozoa / ml are placed under mineral oil in 35 mm multi well petridishes.
2. One or two oocytes are added to each droplet (atleast 20 oocytes are used to estimate the sperm zona binding of each sample).
3. The gametes are incubated in CO2 incubator for four hours.
4. After incubation, the oocyte-sperm complexes are rinsed five times with PBS, using a narrow-bore Pasteur pipette to remove loosely attached spermatozoa.
5. The sperm oocyte complexes are fixed with 2-5% gluteraldehyde in 10 minutes, washed with PBS, stained with 1 mg/ml 'Hochest 33342 dye (sigma) for 10 minutes and washed with PBS.
6. The sperm oocyte complexes are placed on slide slightly compressed with a coverslip and sealed by paraffin/ Vaseline.

Observation

The number of spermatozoa bound to the zona pellucida of the oocytes is counted under fluorescence microscope.

Inference

If more than 200 sperms are attached, the sample is considered to have good fertility

14.4 Heterospermic Insemination and Competitive Fertilization

This test is done to assess the relative fertilizing ability of the spermatozoan from two different males. The pooling of spermatozoa from two different male is done prior to insemination or incubation with oocyte invitro. In this test identification of the source of each spermatozoa is essential. This may be achieved by the use of genetic markers or labeling of the spermatozoa.

14.5 Hemizona Assay

Hemizona assay is a modification of the zona free hamster egg penetration test and heterospermic insemination and competitive fertilization. In this test spermatozoa from different bulls are incubated with hemizona from a single oocyte. A significant relationship exists between the number of bound spermatozoa of a bull and the probability of pregnancy resulting from insemination with that bull's semen.

14.6 Chromatin Analysis

Subfertility can be related to defects of the chromatin, involving increased susceptibility to DNA denaturation in situ. The etiology of this type of defect may be induced by various toxic agents, or there could be a genetic predisposition. The amount and structure of chromatin or DNA has been investigated as a possible means of assessing spermatozoan viability. The assay is based on the estimation of the susceptibility of sperm nuclear DNA to acid denaturation in situ, and the test is called sperm chromatin structure assay (SCSA). The degree of denaturation is determined by flow cytometry measuring the ratio of green (native, double-stranded DNA) and red (denatured, single-stranded DNA) fluorescence emitted from acid treated spermatozoa stained with acridine orange.

14.7 Oviductal Epithelial Cell Explant Test

Cytofluorescent assay can be used to test the reaction of spermatozoa to oviductal epithelial cell monolayers. There is high correlation of spermatozoon morphology and the effect of spermatozoa on oviductal epithelial cell activity, including protein secretion. The number of spermatozoa bound to the explants may also be used as an indicator of spermatozoon viability. However, there is apparent variation in the number of spermatozoa bound depending on the place from which the explant was extracted and the stage of animal's estrous cycle. Oviductal epithelial cell explant test can be used as a potential test for in vitro fertilizing potential of spermatozoa.

Chapter 15

Assessment of Metabolic Activity of Semen

Semen metabolic activity can be assessed with the help of the following tests:

15.1 Incubation test
15.2 Methylene Blue Reduction Test
15.3 Resazurine Reduction Test
15.4 Millovonoo's Resistance Test ('r' test)
15.5 Fructolysis Index
15.6 Oxygen utilization test
15.7 Glutamic oxaloacetic transaminase activity

15.1 Incubation Test

The longevity and viability of spermatozoa during storage is a good index of fertilizing ability of sperm. In this test assessment of motility is carried out over a period time to obtain an idea about the length of time over which acceptable level of motility is maintained in the semen sample. Therefore, this test serves as a mean of determining storage probability within a fairly short period of time. This test is normally carried out at 37°C in bovine semen. But the test can also be performed at 22°C and 5°C. The temperature of the test and dilution rates affect the longevity, therefore, they should be properly standardized.

Materials Required

1. Phase contrast microscope with stage biotherm
2. Water bath
3. Sugar tube
4. Microslide
5. Coverslip
6. Scissors

Test Procedure

1. Thaw a semen straw at 37°C for 30 seconds.
2. Record the progressive sperm motility immediately after thawing.
3. Continue incubation of the semen samples at 37 °C in the water bath and keep on recording the motility every 15 minutes for the first one hour and subsequently hourly till the motility is completely lost.
4. Samples retaining motility for longer period of time are considered as better in terms of viability.

Observation

Per cent motility is recorded under 100 X magnification of a phase contrast microscope after every 15 minutes of incubation for first one hour and subsequently hourly.

15.2 Methylene Blue Reduction Test (MBRT)

The reduction of methylene blue depends on the dehydrogenase activity of the semen. It is based on the principle that the color of the dye is changed from blue to white colorless due to dehydrogenase activity of semen. Hydrogen ions are released in semen due to sperm metabolic activity, which are transferred under anaerobic condition to methylene blue dye and causes change of color of the dye. More is the number of metabolically active spermatozoa in semen, more is the number of hydrogen ions liberated in per unit of time, and less is the time required for reduction of the dye and color change. The test is generally used for neat undiluted semen as the number of spermatozoa/ ml of semen are vital for the time taken for the reduction of methylene blue. The test is very simple but crude test of the metabolic activity of the semen. The limitation of the test is that it cannot be used for frozen semen samples as the number of spermatozoa is vital factor determining the results of the test. The time for reduction of methylene blue can also be influenced by other reducing substances present in the semen or extender like sugar, ascorbic acid and bacteria may also affect the reduction time of methylene blue.

Materials Required

1. Water bath
2. Test tube (10 ml)
3. 0.1, 0.2 and 0.8 ml pipette or micropipette

Chemicals/ Reagents Required

1. Mineral oil
2. Methylene blue solution
3. Egg yolk citrate extender
4. 3 % Sodium citrate

Methylene Blue Solution

2.9 % Sodium citrate Buffer	
Sodium citrate dehydrate2.90 g	
Distilled water upto	100.00 ml

Egg yolk citrate Extender

2.9 % Sodium citrate buffer	80.00 ml
Egg yolk	20.00 ml

Methylene Blue Solution

Methylene Blue	50.00 mg
2.9 % Sodium citrate up to	100.00 ml

Test Procedure

1. Dilute 0.2 ml of neat semen with 0.8 ml egg yolk-citrate extender in a 10 ml test tube.

2. Mix thoroughly
3. Add 0.1 ml Methylene Blue solution and mix.
4. Layer this mixture with ½ inch of mineral oil gently.
5. Place the tube in water bath at a temperature of 46.5°C.

Observation

Start recording the time for the discoloration of the blue color of the dye as soon as the mixture is placed in the water bath.

Inference

Draw Interpretation as follows:

Time required for discoloration of the dye	Semen Quality
3-6 minutes	Good
6-9 minutes	Fair
> 9 minute	Poor

15.3 Resazurine Reduction Test

Resazurine reduction test is also based on metabolic activity of spermatozoa in semen. It is based on the principle that the color of the dye is changed from blue to pink initially and finally it becomes colorless due to dehydrogenase activity of semen. Hydrogen ions are released in semen due to sperm metabolic activity. These ions are transferred under anaerobic condition to resazurine and causes change of color of the dye. More is the number of metabolically active spermatozoa in semen, more is the number of hydrogen ions liberated in per unit of time, and less is the time required for reduction of the dye and color change. The test is generally used for neat undiluted semen as the number of spermatozoa/ ml of semen are vital for the time taken for the reduction of resazurine. Erb et al., 1952 later modified the test by standardizing the concentration of spermatozoa at 150 million spermatozoa in 0.2 ml of diluted semen.

Material Required

1. Water bath
2. Sugar tube/ Screw capped vial
3. Scissors

Chemicals/ Reagents Required

1. Mineral oil
2. Resazurine solution (11 mg resazurine in 200 ml distilled water)

Test Procedure

1. Take 0.2 ml of freshly collected semen in a sugar tube.
2. Add 0.1 ml of resazurine solution to it.
2. Mix the solution gently.
3. Layer the mixture with 1 cm mineral oil. It is done to provide anaerobic conditions required for the reaction.
4. Incubate at 45°C.
5. Observe for the color change.

Observation

Note the time taken for the blue color of resazurine to pink and then from pink to colorless.

Inference

Good quality semen should take approximately 1 minute for the change of blue color to pink and approximately 4 minutes for the pink color to become colorless.

15.4 Millovonoo's Resistance Test ('r' test)

'R' test of semen signifies the ability of spermatozoa to withstand 1 % sodium chloride solution. 'R' value of semen is the volume (in ml) of 1 % sodium chloride solution required to completely cease the progressive motility of all spermatozoa in 0.02 ml of semen. A good quality semen should not have 'R' value less than 5000.

Materials Required

1. Phase contrast microscope
2. Burette
3. 200 ml conical flask
4. 0.02 ml pipettes or micropipettes

Chemical/ Reagents Required

1. 1 % Sodium Chloride

Test Procedure

1. Take 0.02 ml of semen in 200 ml capacity conical flask.
2. Take 1 % Sodium chloride solution in a burrette.
3. Pipette out 10 ml volumes of 1 % Sodium coloride into the flask containing semen.
4. Examine a drop of mixture of semen and Sodium chloride for motility.
5. Keep on repeating addition of 10 ml sodium chloride and motility evaluation till the progressive motility of spermatozoa is completely ceased.

Observation

Observe the motility after each addition of 10 ml of 1 % Sodium chloride under 20 X objective of a phase contrast microscope.

Millovonoo's Resistance ('R' value) is represented as the milliliters of 1 % Sodium chloride solution required to stop the progressive motility of spermatozoa in 0.02 ml of semen. It is calculated as follows:

$$\text{'R' value} = \frac{\text{ml of Sodium chloride solution required}}{0.02}$$

Inference

A good quality semen sample should not have 'R' value less than 5000.

15.5 Fructolysis Index

Fructolysis index can be defined as the milligram of fructose utilized by 10^9 spermatozoa in 1 hour at 37 °C. Greater is the metabolic activity of the spermatozoa more will be the amount of fructose metabolized per unit of time in the semen sample.

Materials Required

1. Test tubes
2. Micropippette (variable volume)
3. Water bath (37 & 80°C)
4. Electric colorimeter/ Spectrophotometer

Chemicals/ Reagents Required

1. Fructose
2. M/4 Phosphate buffer
3. 2 % Zinc sulphate
4. N/10 NaOH
5. 0.1 % resorcinol
6. Ethanol
7. 30 % hydrochlorioc acid

Standard Fructose Solution

Fructose	400 ml
Standard benzoic acid	100 ml

Prepare 0.02% and 0.1 % fructose solution from the stock

Blank Solution

2 % Zinc sulphate	0.5 ml
Sodium hydroxide solutions	0.5 ml
Standard fructose solution	1 ml

Phosphate Buffer

3.4 % potassium dihydrogen phosphate	9.6 ml
3.55 % sodium hydrogen phosphate solution (pH 7.4)	40.4 ml

Test Procedure

1. Pippette 0.4 ml of freshly collected semen of known concentration into a test tube containing 0.6 ml of M/4 phosphate buffer (pH 7.47).
2. Add 0.1 ml of semen buffer mixture to 1.9 ml of distilled water in a test tube.
3. Add 1 ml of 2 % Zinc sulphate solution and 1 ml of N/10 sodium hydroxide solution to above contents. Deproteinise it (0 hr sample)
4. 0.9 ml of semen buffer mixture (Step 1) is incubated immediately at 37°C for one hr.

5. After incubation deprotenise it as per method in Steps 2 and 3 (1 hr sample)
6. Heat both the deproteinised samples in a boiling water bath for one minute.
7. Cool and filter the samples.
8. Transfer 0.5 ml of each filtrate to individual test tubes.
9. Make the volume to 2 ml with distilled water.
10. Add 2 ml of 0.1 % resorcinol in ethanol and 6 ml of 30 % hydrochloric acid to each sample.
11. Shake the contents and maintain the samples in a water bath at 80 °C for 10 minutes.
12. Immediately cool the contents to room temperature in running water.
13. Chocolate brown color develops in the solution.
14. Compare the intensity of color with the help of a visual electric colorimeter with a standard fructose solution of equal volume.

$$\text{Concentration of Unknown sample} = \frac{\text{Reading of standard X Concentration of standard}}{\text{Reading of unknown}}$$

With this method amount of fructose present in semen immediately after collection and after 1 hr incubation are worked out. The difference between the concentrations of fructose in 1 hr is on account of fructolysis in 1 hr.

Fructolysis may be worked out in respect of 10^9 live sperm cells. This method is useful as compared to 10^9 sperm concentration, and provides a true comparison of metabolic activity of semen sample. Since fructolysis method is not rapid test, its usefulness in semen evaluation is not of much concern.

15.6 Oxygen Utilization Test

Oxygen is utilized for the oxidation of the substrates outside cell wall (exogenous respiration) and also for the oxidation of intracellular material (endogenous respiration). If the spermatozoa are very active, they utilize more oxygen per unit of time. The assessment of the oxygen uptake by the spermatozoa is done by Warburg's apparatus. The sperm cells are diluted in a suitable dilutor and are kept in a small flask, at 37°C in a water bath. This is a closed system, wherein the CO_2 produced is absorbed by an alkali. The oxygen consumed consumed and CO_2 produced can thus be estimated.

15.7 Glutamic Oxaloacetic Transaminase Activity

Glutamic oxaloacetic transaminase is a purely cellular enzyme and is not normally found in the seminal plasma. The presence of this enzyme in the seminal plasma indicates damage to the spermatozoa. For estimation 1 ml substrate is taken in a test tube and placed in a water bath maintained at 37°C. Add 0.2 ml of extra cellular fluid (Sample). Keep in the water bath at 37°C for exactly 1 hour. After 1 hour add 1 ml of color reagent to stop the enzyme activity and start color reaction. The sample is then left at room temperature for 20 minutes. After 10 minutes add 10 ml of 0.4 N NaOH and mix by inversion after putting rubber stopper. After 5 minutes the color is read in a double cell colorimeter with green filter. Every estimate is accompanied with a blank using 1 ml substrate, 0.2 ml distilled water, 1 ml color reagent and 10 ml 0.4 N NaOH. Corrections are made from readings of blank every time before calculation.

Chapter 16

Miscellaneous Tests for Evaluation of Semen Quality

The following tests are discussed under this section:

16.1 Estimation of ascorbic acid concentration
16.2 Effect of cold shock
16.3 Mitochondrial activity
16.4 Capacitation of spermatozoa and calcium influx
16.5 Morphometric analysis
16.6 Flow cytometry

16.1 Estimation of Ascorbic Acid Concentration

Ascorbic acid concentration in semen is beneficial for its antioxidant action. It protects the spermatozoa against oxidative damage of free radical produced in the semen.

Materials Required

1. Test tubes with stand
2. Burrette with stand
3. Centrifuge tube with stand

Chemicals/ Reagents Required

1. Standard Ascorbic acid solution (1.2 mg per 100 ml)
2. Dye (2,6-dichlorophenol indo phenol)

Standard Ascorbic Acid Solution

Stock Solution

Ascorbic acid	60.0 mg
5 % Acetic acid up to	100.0 ml

Working Solution

Stock solution	1.0 ml
5 % acetic acid up to	100.0 ml

Preparation of the Dye (2,6-dichlorophenol indo phenol)

1. Extract 100 mg of 2,6-dichlorophenol indo phenol through a filter paper with 25 ml of boiling distilled water (Stock solution). Keep the solution in refrigerator.
2. Dilute 5.0 ml of stock solution to 100 ml with freshly boiled and cooled distilled water.

Estimation Procedure

1. Add 0.8 ml 15 % trichlor acetic acid to 0.2 ml of freshly collected semen. Mix the contents and filter it through filter paper no. 1 (Sample).
2. Add 0.05 ml of 2,6-dichlorophenol indophenol solution in a centrifuge tube, and titrate it against the protein free filtrate of semen in a micropipette.
3. Observe for the pink color to disappear. Record the end point.
4. Titrate a blank solution of 0.2 ml standard ascorbic acid and 0.8 ml of 10 % trichlor acetic acid (Standard) against 0.05 ml of 2,6-dichlorophenol indo phenol dye.

Observation

Record the end points of titration of semen solution (sample) and standard ascorbic acid solution (standard) and calculate the ascorbic acid concentration as per the following equation:

$$\text{Ascorbic acid concentration (mg/ 100 ml)} = \frac{\text{Titration reading of standard X 1.2}}{\text{Titration reading of the sample}}$$

16.2 Effect of Cold Shock

This test may be useful to grade semen for varying proportion of cold shock resistant spermatozoa. It is likely semen with higher concentration of cold shock resistant spermatozoa keeps longer under preservation and may give higher fertility. This test may be used to get a rough indication of freezing capacity spermatozoa. More is the number of cold shock resistant spermatozoa in the semen sample better will be freezing ability of the sample. The cold shock resistance varies between individuals, breeds and species.

Materials Required

1. Microscope with 100 X oil immersion objective
2. Refrigerator/ ice bath
3. Microslides
4. Test tubes
5. Sugar tubes
6. Test tube stand

Chemicals/ Reagent Required

1. Eosin-nigrosin stain

Hancock's (1951) Stain

Eosin Y	5.00 g
Nigrosin	30.00 g
Distilled water	300.00 ml

Dissolve nigrosin in distilled water while stirring and heating until it is dissolved. Later dissolve eosin in the nigrosin solution in a similar manner. Do not boil.

Campbell et al (1960) Stain

Nigrosin Solution	150.00 ml
Eosin Y (GT Gurr)	5.00 g
Stock Buffer Solution	30.00 ml

Contd...

Contd...

Stock Glucose Solution	30.00 ml
Distilled Water up to	300.00 ml

Dissolve by stirring and heating over a hot plate. Do not boil. Filter and store at 4-5 °C.

Test Procedure

Method 1

1. Calculate live per cent of the sample to be evaluated by eosin-nigrosin method.
2. Draw a small quantity of semen from the sample and keep at 0°C for 10 minutes.
3. Determine per cent live sperms surviving the cold shock. The percentage of sperms remaining alive after the shock may be expected to maintain fertility for a longer time under proper storage conditions.

Method 2

1. Take ten drops of eosin nigrosin in a small test tube.
2. Maintain it at 0 °C.
3. Take freshly collected semen and evaluate for per cent live with eosin-nigrosin stain.
4. Add one drop of freshly collected semen to chilled stain and keep it at 0 °C for 15 minutes.
5. Prepare a smear and evaluate live sperm per cent.

Method 3: LT 50 Method

It represents the temperature at which semen sample is chilled to bring the number of live sperms to half.

16.3 Mitochondrial Activity

Mitochondria are located in the middle piece of the spermatozoa, and they contain enzymes and cofactors necessary for the production of ATP which is essential for the sperm to maintain its motility. Mitochondrial activity is evaluated by Rhodamine 123 (Rh123) that is accumulated primarily by the mitochondria within the spermatozoa. R123 is a cationic fluorescent molecule and its fluorescence is emitted only from functioning mitochondria. The intensity of Rh123 fluorescence depends on the total amount of functional mitochondria in the flagellum. This staining method is usually used together with one of the supravital probes.

16.4 Capacitation of Spermatozoa and Calcium Influx

Capacitation is defined as all the physical and biological changes in the spermatozoon, which enable the sperm cell to undergo the acrosome reaction. During this process, calcium enters the sperm cell. Ionophore A23187 and heparin treatment can induce capacitation leading to the acrosome reaction, characterized by calcium influx into the cell. This influx is measured by flow cytometry using fluorescent probe fluo3- AM.

16.5 Morphometric Analysis

Morphometric analysis is used for measuring the length, width, area, perimeter and nuclear density of the sperm head performed on stained smears, either manually with an ocular-mounted micrometer or by computer assisted microscopy. For staining, Papanicolau or Feulgen stains are used.

16.6 Flow Cytometry

A correlation between optimally functioning mitochondria and viable spermatozoa has been reported in the past. Therefore, utility of flow cytometry with both single and double staining to assess the functional capabilities of spermatozoan mitochondria has been explored and it promises useful conclusion in precise semen evaluation.

Chapter 17

Estimation of Bacterial Load in Semen Samples

The bacterial load in the semen samples is estimated by standard plate count method. The bacterial contaminants of semen have been a major concern for most of the frozen semen production laboratories as it adversely affects semen quality and hence subsequent fertility. There exists a significant negative correlation of bacterial load and sperm livability, motility and in vitro fertility tests. Macrophages and polymorphonuclear granulocytes, which form the first line of defense against these microorganisms, produce reactive oxygen species (ROS) to kill these microorganisms. ROS, however, is also released outside these cells and may react with molecules and cells such as spermatozoa in their vicinity. The cellular antioxidants, present mostly in the cytoplasm, are scanty and inadequate to counteract this ROS as the sperm cell cytoplasm is very small (20 μm^3) and mostly distributed in the midpiece. Therefore, semen samples with minimum bacterial load are expect to give better results on insemination.

The freshly ejaculated semen should be free of microorganisms unless the testis and accessory sex glands are infected. Apart from the infected testes and accessory sex glands the microbes may also be added to the semen through urine, prepucial cavity, the prepucial orifice or by blood or tissue fluid extravasated into the urogenital sytem.

Few bacteria of semen survive at –196°C in liquid Nitrogen and acquire a certain level of resistance to antibiotics and account for the contamination of approximately 50% of frozen semen samples. The bacterial contaminants of the semen have been classified as pathogenic, potentially pathogenic or non-pathogenic.

Materials Required for Estimation Bacterial Load

1. Vertical laminar flow
2. Scissors
3. Micropitte of 1000 μ
4. Petridishes of 90 mm diameter
5. Spirit lamp
6. Methanol
7. Autoclave
8. Water bath maintained at 45°C
9. Water bath maintained at 37°C
10. Vortex shaker
11. Bacteriological Incubator

Chemicals/ Reagents Required

1. Sterile Normal saline (0.85%)
2. Agar Medium

Normal Saline Solution

850 mg (0.85%) sodium chloride in 100 ml distilled water

Agar Medium (as per Bureau of Indian Standards, 1991)

Meat Extract	4.0 g
Anhydrous D Glucose ($C_6H_{12}O_6$)	0.4 g
Dehydrated Yeast Extract	1.0 g
Bacto-Peptone	1.2 g

Contd...

Contd...

Sodium Chloride	0.8 g
Disodium Hydrogen orthophosphate dodecahydrate ($Na_2HPO_4.12H_2O$)	0.8 g
Gelatin	4.0 g
Agar Powder	5.6 g
Distilled Water up to	400 ml

Dissolve the components in distilled water by boiling. Check the pH and adjust the pH, if required, to ensure a pH of 7.4±0.1 after sterilization. Sterilize the media by autoclaving at 15 psi pressure at 121± 1 oC for 20 minutes. Cool the media to 45 ±0.5 oC in a water bath before use. Enrich the medium by adding 10 % inactivated bovine serum sterilized by ultrafilteration or tyndallization at the time of use.

Readymade preparations from companies like Himedia, Qualigens and SRL are also available which can be used as per manufacturer's instructions.

Estimation Procedure

1. Prepare diluent by adding 850 mg (0.85%) sodium chloride in 100 ml distilled water. Distribute 9 ml diluent in 50 ml glass tubes, plug it with cotton and autoclave at 15 lb for half an hour. Prepare two such tubes for each sample to be analysed.
2. Thaw three frozen semen straws (0.5 ml) at 37°C for 30-60 seconds.
3. Evacuate the contents of the straw in a sterile sugar tube under a vertical laminar air flow.
4. Take 1 ml and pour in 1st tube containing 9 ml diluent.
5. Mix thoroughly by vortex shaker.
6. Take 1 ml from first tube and pour it in second tube containing 9 ml diluent.
7. After thorough mixing take 1 ml of diluted semen from each tube in separate sterilized Petri dishes.
8. Pour about 15-20 ml of agar media in each petridish.
9. Mix by rotating gently on the working surface of the laminar flow.
10. Allow medium to solidify and then incubate the plates in inverted position for 72 hr at 37°C in a bacteriological incubator.

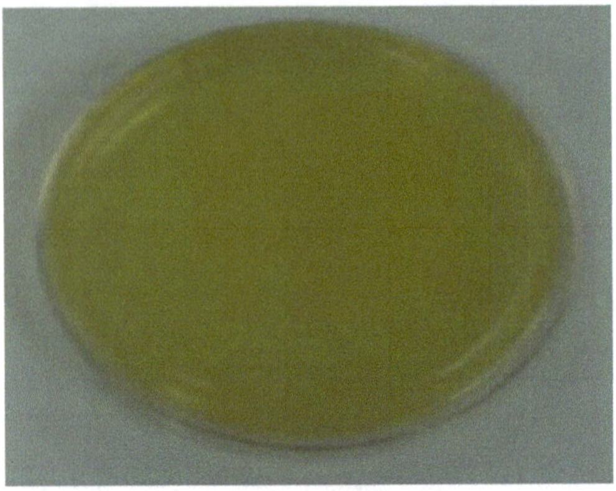

Control (plate without any bacterial colonies)

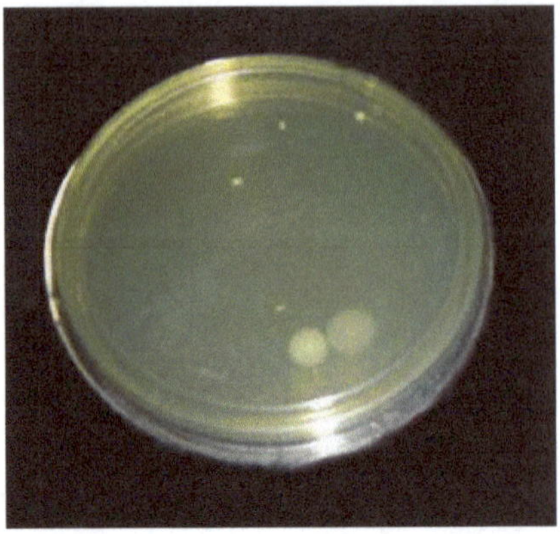

Plate containing bacterial colonies

Standard plate count for estimation of bacterial load in semen

Observation

After the end of incubation period count only well distinguishable colonies with the help of colony counter or manually. Ignore the colonies that have grown on the surface of the plates. Multiply number of colonies in each plate by their respective dilution to get the number of Colony Forming Units (CFU) per ml. Take the average of the two plates of each sample. And express the results as average CFU ml-1.

Note

1. If even a singly colony is found in the control after the end of the incubation period, the whole procedure has to be repeated.
2. If any plate contains more than 300 colonies, the procedure has to be repeated for that particular sample with higher dilutions (1:1000 and 1:10000).

Inference

BIS recommends that bacterial load less than 500 CFU ml-1 of frozen semen is acceptable for AI.

Section D

Artificial Insemination

Chapter 18

Estrus Detection and Preparation of Teaser Bull

18.1 Estrus Detection

18.1.1 Estrus Detection on the Basis of External Signs

The following signs are suggestive estrous period in cattle and buffaloes

- Partial inappetance due to excitement
- Drop in milk yield
- Restless ness of varying degree
- Frequent micturition with raised tail and crouching the back and lumbar region
- Marked bellowing
- Swollen vulvar lips- Due to swelling the wrinkles on vulvar lips disappear. The vulvar lips appear turgid and stand prominently.
- Congestion of vulvar lips
- A clear, shiny mucus discharge seen at the vulvar lips extending down to perineal region and smearing the hind quarters. The discharge is stingy, having good consistency and no odour.

- Homosexual behaviour- The cow mounts on other cows and stands to be mounted.
- Standing to be Mounted is the Most Definite Sign of Estrus
- Vulvar lips of the cows in heat are sniffed frequently by the other cows.

18.1.2 Estrus Detection on the Basis of Findings of per Rectal Examination

- Cervix is relaxed
- External os cervix is open
- Uterine horns are tonous
- Fully grown follicle is present on one of the ovaries

18.1.3 Estrus Detection using a Teaser Bull

In organized farms vasectomised bulls or androgenized cows are paraded in the herd twice daily preferably during cooler parts of the day i.e. early morning and late evening. The cows which stand to be mounted by the teaser are in estrus. This is the most efficient method of estrus detection in organized farms.

Aids to Improve Estrus Detection Efficiency in Cattle and Buffaloes

- Tail paint, when applied to the base of tail and sacrum, is removed by rubbing when a cow stands to be ridden. It is not specific but it is cheap and quite effective when used sensibly on a selective basis.
- KaMaR heat mount detectors are activated in the same way as that described for tail paint. They are more expensive than the later and cows must be identified when they are affixed because in some cases riding displaces the device.
- Closed circuit television with time lapse video is expensive to install but is quite effective when used selectively, for example during hours of night when cows are not observed.
- Teaser bulls or androgenised cows will identify the cows which are in estrus. They should be provided with some form of marker such as chin ball device.

- Measurements of certain physiological changes such as increased body temperature, alterations in electrical impedence within vagina or vaginal mucus can be used but it requires specific instruments.
- Study of fern pattern in cervical mucus can be helpful in predicting estrus and time of breeding. This can be done with the help of commercially available crystoscope.
- Hormonal assay especially estrogen / progesterone can be helpful.

18.2 Preparation of Teaser Bull

Teaser bulls are used to detect cows in estrous. They can be prepared by any one of the following ways.

18.2.1 Vasectomy

18.2.2 Caudectomy of epididymis (Epididymotomy)

18.2.1 Vasectomy

Vasectomy can be performed at any one of the following sites:

1. Posterior side of the scrotal neck, little lateral to the median line.
2. Anterior side of the scrotal neck on either side of median line.
3. Lateral aspect of the scrotal neck.

Technique of Vasectomy

The skin and tunica dartos are incised and spermatic cord enclosed in tunica vaginalis is pulled out through the incision site with a blunt instrument. An incision is made on the tunica vaginalis and the vas deference is traced. Clamps the vas deference with an artery forceps. Put ligature on either side of the artery forceps. Remove the piece of the vas deference between the ligatures by separating it from attachment of the visceral layer of tunica vaginalis. The portion of vas deference left back is not separated from its attachment of tunica vaginalis to safe guard against reunion. Suture the scrotal incision. The operation is repeated on the other side also. Semen is collected and examined for the presence of spermatozoa once a week. By the end of third week semen samples should be completely free of spermatozoa.

18.2.2 Caudectomy of Epididymis (Epididymotomy)

The tail of the epididymis at the bottom of the scrotum is held tense against the skin and a small incision is made incising the skin, dartos and tunica vaginalis. The tail of the epididymis is exposed excised off near the lower portion of the testicles. The same operation is repeated on the other side. Skin sutures need not be made unless the incision is large enough to cause prolapse. There is possibility of re-canalization of the epididymis since ligature is not made. Therefore, modification of this technique is to expose the epididymis little more and then ligature either portion of its bend before snipping it off distal to the ligature.

Semen is collected and examined for the presence of spermatozoa once a week. By the end of third week semen samples should be completely free of spermatozoa. It should further be retested every three months to prevent possible re-canalization of the vas deference.

Chapter 19

Thawing of Frozen Semen

The process of converting the frozen semen to liquid state while retaining maximum semen quality is called thawing. The process of thawing is as important as freezing as there is equal probability of loss of semen quality in freezing and thawing. Therefore, great care should be taken while thawing to insure that it is done with minimum loss of sperm quality. There are two methods of thawing:

- Rapid thawing
- Slow thawing

The effectiveness of each method of thawing depends upon the extender composition and freezing procedure. With any thawing procedure, the recommendations of the semen supplier should be followed. The thumb rule is that rapid freezing rates require rapid thawing. Regardless of the thawing procedure, sperm injury can occur if semen is mishandled post-thaw. Straws thawed in warm water should be dried thoroughly to avoid the possibility of water contact with the semen when the straw is opened; fluctuations in the temperature of thawed semen should be avoided to minimize the risk of chilling injury; and the interval from thawing to insemination should be kept to a minimum because post-thaw survival of sperm is short in the straw.

Procedure

Identify the canister in the cryocan with desired semen straw. Lift the canister upto the neck of the cryocan. Pick up the desired straw from the cryocan with a precooled forceps. Shake the straw gently in air for 1-2 seconds. Subsequently any one of the following should be done immediately:

1. Dip the straw in water bath with a stirrer maintained at 70°C for 5 seconds (Rapid thawing).
2. Dip the straw in water bath with a stirrer maintained at 35-37°C for 30-60 seconds (Slow thawing).

Take out the straw. Wipe it dry. Hold the straw at the laboratory end. Shake it with a jerk to create the air space at the laboratory end. Cut the straw at right angle at the air space created at the laboratory end. Load in AI gun and proceed for insemination or evacuate the semen into a sterile sugar tube and proceed for semen evaluation.

Important Points to be Considered during Thawing

1. Calibrate the thermometer used for measuring the temperature of water used for thawing on a regular basis to avoid temperature variation.
2. Never thaw the semen in your pocket or by rubbing in your palm as in either of these procedure the thawing temperature is not maintained and the thawing is too slow and it will lead to reduction in the survivability of viable sperms.
3. Completely dry the semen straw after thawing with a paper towel as water is lethal to sperm cells.
4. The number of straws thawed should not exceed the quantity that can be used within next 5-10 minutes.
5. The quantity of water used for thawing should be sufficient for dipping the whole straw.

The water used for thawing should be agitated during thawing so that the straw is uniformly thawed at the desired temperature. This becomes particularly more important when several straws are thawed at a time so as to avoid the straws sticking and freezing together. This also causes the frozen semen straws to create a cold zone around it and prevent thawing at the desired temperature (37°C).

Chapter 20

History and Technique of Artificial Insemination

Artificial insemination (AI) is the technique of placing sperm into a female's uterus (intrauterine), or cervix (intracervical) using artificial means rather than by natural copulation. Modern techniques for artificial insemination were first developed for the dairy cattle industry to allow many cows to be impregnated with the sperm of a bull with traits for improved milk production. It is the first generation reproductive biotechnology which has revolutionized the animal productivity. It has now been established as cheapest and fasted mode of genetic improvement of livestock. AI can be done using fresh semen, semen preserved at room temperature, refrigerated (4 °C) and cryopreserved (-196 °C) semen. The use of AI with frozen semen has become very popular and is being successfully used throughout the country.

20.1 History of Artificial Insemination (AI)

Many people think of artificial insemination as a modern technology but it has a long history. According to the history, artificial insemination was attempted on Juana, wife of King Henry IV of Castile.

The history of AI is interesting. Old Arabian documents dated around 1322 A.D. indicate that an Arab chieftain wanted to mate his prize mare to an outstanding stallion owned by an enemy. He introduced a wand of cotton into the mare's reproductive tract, and then used it to sexually excite the stallion causing him to ejaculate. The semen was introduced into the mare resulting in conception.

In 1677 the Dutch scientist Anton van Leeuwenhoek saw spermatozoa through the newly invented microscope. This finding led to further research. Lazzaro Spallanzani is usually considered the inventor of AI. His scientific reports of 1780 indicate successful use of AI in dogs. With this new knowledge, Spallanzani experimented on frogs, fish, and other animals and was successful.

In 1899, Ivanoff of Russia pioneered AI research in birds, horses, cattle and sheep. He was apparently the first to successfully inseminate cattle artificially. Mass breeding of cows via AI was first accomplished in Russia, where 19,800 cows were bred in 1931. Denmark was first to establish an AI cooperative association in 1936. E.J. Perry of New Jersey visited the AI facilities in Denmark and established the first United States AI cooperative in 1938 at the New Jersey State College of Agriculture.

Efforts to develop practical methods for AI were started in Russia in 1899. Papers on artificial insemination in horses had been published by 1922. By the mid 1940s artificial insemination had become an established industry. In 1949 improved methods of freezing and thawing sperm were developed. The idea for adding antibiotics to the sperm solution came in 1950 from Cornell. Improved methods of sperm collection were developed in the 1970s and 1980s. Research to improve methods of artificial insemination continues and is now routinely studied under animal science/ reproduction curricula.

History of Artificial Insemination: At a glance

- Arab Chieftains - Stole semen to breed mares
- Leeuwenhook (1677) - Used microscope to see sperm
- Spallanzani (1780)
    - Sperm could fertilize
    - Cooling and freezing inactivated sperm and upon warming sperm were reactivated
- Ivanov of Russia (1900)

- Developed methods as we know today
- Most work was with horses but did some work in cattle and pig

❑ Denmark (1933)

- First dairy cooperative

❑ First US AI Cooperative (1937)

- First US dairy cooperative in New Jersey

❑ First Indian to perform AI: Sampat Kumaran at Mysore Dairy Farm in 1939.

Advantages of AI

1. *Cost of rearing the bull in each and every farm can be avoided* : Normally one bull is required for around twenty five breedable cows for natural service and has relatively short life (2 years) in the herd.
2. *Maximum utilization of superior germplasm* : The technique of semen collection, dilution, cryopreservation and AI has tremendously increased the utilization of superior germplasm. One ejaculate which can other wise be used to impregnate one cow can now be diluted and can be used to produce approximately 500 doses for AI.
3. *Prevents the spread of veneral disease* : If a bull used for natural service picks up a veneral disease, it can spread it to the whole herd bred with that bull, which can be avoided if healthy bulls are selected for semen production and AI programme.
4. *Prevents the spread of infectious and genetic diseases* : Strict Government norms for frozen semen production insures the selection and maintenance of bulls free of infectious diseases like TB, Johns disease, Brucellosis, IBR, BVD, campylobacteriosis and Trichomoniasis and genetic diseases like Bovine leukocyte adhesion deficiency syndrome (BLAD), Citrullinemia, diuridine monophosphate deficiency syndrome (DUMP), Factor XI deficiency and cytogenetic disorders. Thus extensive use of artificial insemination from semen of bulls maintained in organized semen station can prevent the spread of these infectious and genetic diseases in the population.

5. *Fastest mode of genetic improvement* : Superior germ plasm can be propagated at fastest and most economic way by artificial insemination.
6. *Easy transport of germplasm* : The male germplasm can be transported across the states and countries in most economic way.
7. It permits the breeding from a bull which is unable to serve because of injury or in case the bull is too heavy for a particular cow.
8. AI coupled with semen cryopreservation has enabled prolonged storage of desired germplasm and its use in future when required even after the death of the bull.
9. Progeny testing has been made easy by the use of AI as sufficient number of daughters can be produced in a limited time.
10. Desired semen can be used to inseminate a particular cow and hence it avoids the risk of inbreeding.

Limitations of AI

1. It requires skilled technicians with proper knowledge of palpation of female genitalia, semen handling and AI technique.
2. Precise estrus detection is required for desired results with AI.
3. Strict hygienic conditions and sterilization should be maintained while inseminating animals.

Techniques of Artificial Insemination (AI)

The technique of artificial insemination was discovered by Lazzaro Spalanzani. The following methods are nowadays used for insemination in farm animals:

1. Cervical Insemination by use of vaginal speculum- Sheep and goat
2. Vaginal method - Mare, sheep
3. Recto vaginal method- Cow and buffalo

Vaginal Insemination

This method is the most ancient method of AI. Here semen is deposited in the vagina with the help of catheter or tube. In cattle a tube of approximately 16 inches is used for vaginal insemination. Tip of the tube is pointed upward as it enters vagina to avoid catching in the sub urethral diverticulum or entering the urethra. The disadvantage of this method is that it requires large volume of semen. The conception rate is very low in inseminations done by this method if small quantity of semen is used. Here highly motile spermatozoa in high concentrations need to be used if reasonably good conception rate is desired. The possible cause of low conception rate in this method is that a large number of the spermatozoa are lost from the vaginal tract before entering the cervix. This method of insemination is used in mares.

Cervical Insemination

In this method the vaginal walls are spread apart with the help of speculum. A well lubricated speculum (2 to 3 cm in diameter and 35 to 40 cm long for cattle) is inserted in such a way that the external os of the cervix is visible against a flashlight worn on the head of the inseminator. A simple torch light can also be used. After locating the external os cervix a loaded AI gun is guided through the external os in the cervix to approximately mid cervix and semen is deposited as deep as possible in cervix. Disadvantage of this method is that it requires repeated sterilization of the speculum before each insemination. The conception rate is also lower as compared to recto vaginal method. This method is commonly followed in sheep and goats where the recto vaginal method cannot be followed.

Recto Vaginal Insemination

This is the most popular technique of AI. In this method a well lubricated gloved hand is inserted into the rectum and the rectum is evacuated by back racking. The cervix is caught hold and pushed back cranially to stretch the vaginal tract to avoid stucking up of catheter in vaginal fold. Thumb is placed on external os cervix. After catching the cervix an AI gun or catheter is inserted into the vagina at an angle of 30° pointing upwards upto a point where it is obstructed and then horizontally. The tip of the catheter should be felt at the thumb on the external os cervix and the catheter is inserted into the cervix. The catheter is gently passed through the cervix by manipulating the cervical rings by changing the direction of the catheter. The catheter should be withdrawn a little as it gets obstructed in the cervical rings and then it is again directed forwards till it reaches the site of semen deposition. The semen is then deposited and the catheter/ AI gun is withdrawn gently followed by withdrawing hand from the rectum. This method requires considerable skilled technician for insemination. The advantage of this method is the higher conception rates as compared to other methods. This method does not require sterilization of any instruments besides introduction of hands per rectum and manipulation of the genitalia also stimulates them as in case of natural service.

Insemination of Cattle and Buffaloes with Frozen Semen by Recto Vaginal Method

Thawing of Frozen Semen Straw

- Identify the canister from which the desired straw is to be taken.
- Remove the lid of the LN2 container, lift the proper canister upto the level of the frost line. Never lift the canister above the frost line.
- Cool the tip of straw holding forceps by holding it under the neck for few seconds. Grasp an individual straw and remove it. At the same time lower the canister immediately back into the container. If you are unable to remove the straw with in 10 seconds by forceps then lower the canister back in to the LN2 container and wait for next few seconds and next attempt. It is not advisable to thaw more than one straw at a time.
- Shake the straw in the air gently to expel LN2 trapped at the end of the factory seal of the straw.
- Dip the straw into clean water bath of 37 °C for 30-60 seconds. During thawing the entire straw should be completely submerged in water bath.
- Remove the straw from water bath, dry the straw with clean towel/ tissue paper// absorbable cotton. Inspect the straw carefully and discard the straws with cracks or defective seal.

Loading of the AI Gun

- Hold the straw at laboratory end. Shake it once to create air space at the laboratory end.
- Before loading the straw in the AI gun, warm the gun by rubbing it with a paper towel to avoid cold shock to the semen. This

process is of particular significance in winter season or even during cooler parts of the day.

- Place the straw in AI gun and cut the laboratory end of the straw at right angle at the air space. Make sure that the cut straw has clean cut edge to avoid back flowing of semen during insemination.
- Fix a sheath over straw and AI gun to get a perfect fit between the extremity of straw and cone of AI sheath. Do not touch the tip or middle part of the sheath to avoid contamination. Only base portion of the sheath should be handled.
- Fix the sheath locking with plastic 'O' lock.
- To ensure better conception rate deposit the thawed semen in to female genitalia as early as possible.

Optimum time and Place of Insemination

The most critical step in insemination of livestock is the delivery of the semen to the right place and at the right time. Because the cow may ovulate from either the left or right ovary, the "right place" for semen deposition is the uterine body where right and left uterine horns join allowing sperm equal access to both sides. Recommendations for the "right time" are based on the duration of estrus (heat), the timing of ovulation, and the lifespan of the sperm and egg. Estrus in the cow lasts for less than a day (an average of 18 hours), ovulation occurs 10-14 hours after the end of estrus, the oocyte survives for 12-14 hours after ovulation, and sperm survive for 24-48 hours post-insemination (although this time may be shorter for frozen thawed semen). The recommendation to breed 12-18 hours after the onset of estrus is based on these figures and results in highest fertility. Presently it is believed that a reservoir of perhaps only a few thousand sperm is established in the oviduct below the site of fertilization prior to the arrival of the egg. Sperm within this reservoir appear to survive and retain their fertilizing ability longer while going through the changes necessary to unite with the egg when it arrives. The optimal timing of insemination may be related to the establishment of this reservoir during late estrus which in turn may be related to differences in bull fertility and optimal sperm number per insemination.

Table: Optimum time of Insemination

Animal	Length of estrous cycle (days)	Length of estrum	Time of ovulation	Optimum time for service
Cow	18-24	10-14 h	10-14 h after end of estrus	Mid to late estrus (12-18 hours after the onset of estrus)
Buffalo	21-22	10-32 h	10-14 h after end of estrus	Later half of estrus
Sheep	14-20	30-36 h	20-30 h after start of estrus	18-24 h after the onset of estrus
Goat	15-24	30-60 h	Last day of estrus	24-26 h after onset of estrus
Mare	19-23	2-11 days	24-28 h before end of estrus	Every alternate day beginning from the second day of estrus
Sow	18-24	1-3 days	30-36 h after start of estrus	12-30 h after onset of estrus
Bitch	1-4 cycles/year	4-13 days	24-36 h after start of estrus	2-3 d after onset of true heat or 10-14 day after onset of post estrus bleeding
Camel	10-21	1-7 days	-	-
Cat	15-21	1-10 days	24-30 h after coitus	-

Insemination Technique

- Work under hygienic conditions.
- Handle frozen semen containers, straw and AI equipment scientifically and correctly.
- Examine the animal for proper stage of estrus before thawing the straw.
- Pass your hand per rectally in the shape of a cone. Catch hold of the cervix. Put your thumb on external os cervix. By that time the assistant will thaw the semen straw and prepare the AI gun.
- Before inserting the AI gun the assistant will wash and wipe the vulvar lips with clean towel and pull apart the vulvar lips with both hands to avoid entry of dung or any contaminant with AI gun in the genital tract.

- Hold the AI gun with only two/ three fingers to avoid the putting pressure during insemination. Insert the AI gun first at 35-45° and then horizontally forward and pass along dorsal side of the vagina. This is to avoid chances of entry of the catheter into the sub urethral diverticulum or into the urethra.
- Manipulate the cervical fold by withdrawing the gun little backward and changing its direction when it gets stuck into the cervical folds. The cervix can also be manipulated instead of changing the direction of the gun. Never apply force while passing AI gun.
- Stop the instrument as soon as it reaches the body of the uterus, and do not withdraw the instrument, especially when the cow urinates. Deposit the semen slowly (for 5 seconds) in the caudal part of the body of the uterus near internal os cervix and withdraw AI Gun care fully.

Insemination Technique with Chilled Semen

The insemination with chilled semen is performed with the help of a glass catheter and syringe with rubber connector. The glass syringe is attached to a rubber connector and semen is sucked from the semen storage tube. The syringe is then attached to a glass pipette through the rubber connector. The left hand is inserted into the rectum and the cervix is grasped. The pipette is then introduced into the vagina by the right hand of the inseminator and directed through the cervix. Semen is deposited in the same manner as artificial insemination with frozen semen by recto vaginal method.

Tips for Beginners

a. Insert the instrument into vagina taking care to avoid to entring suburethral diverticulum or external urethral orifice.

b. When muscular contractions force the reproductive tract toward anus and cause vagina to become folded, grasp the cervix with left hand and push it forward to straighten the vaginal folds.

c. When cow attempt to expel left hand from rectum with peristaltic muscular contractions, wait until the rectum is relaxed.

d. When the vagina fills with air, dispel the air by holding the last rectal fold and pulling it towards the anus to expel the air and simultaneously blocking with the arm, the entry of the air into the rectum.

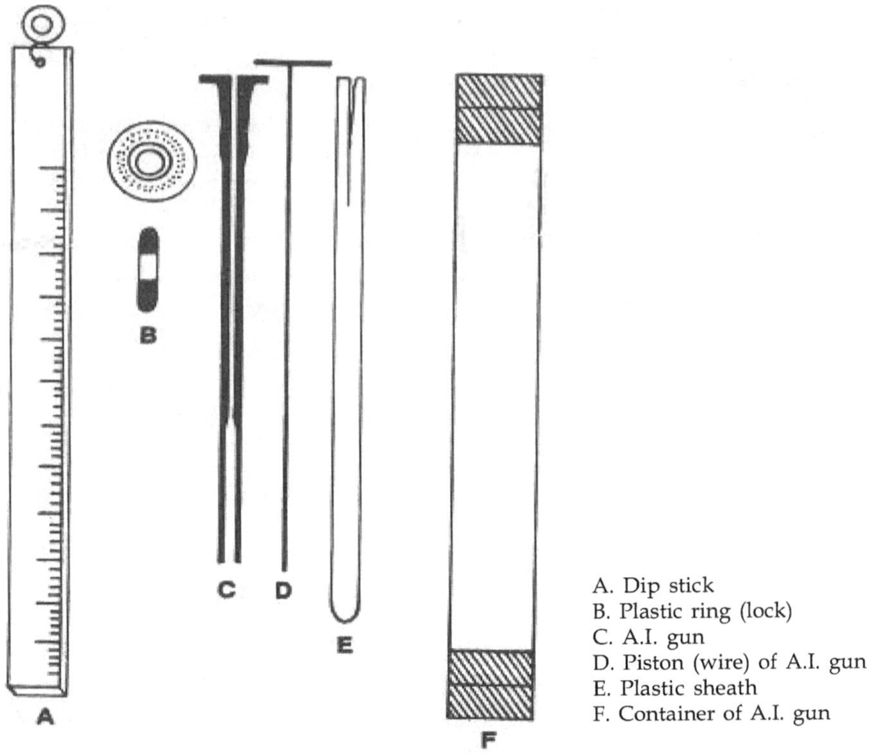

Various instruments used in Artificial Insemination

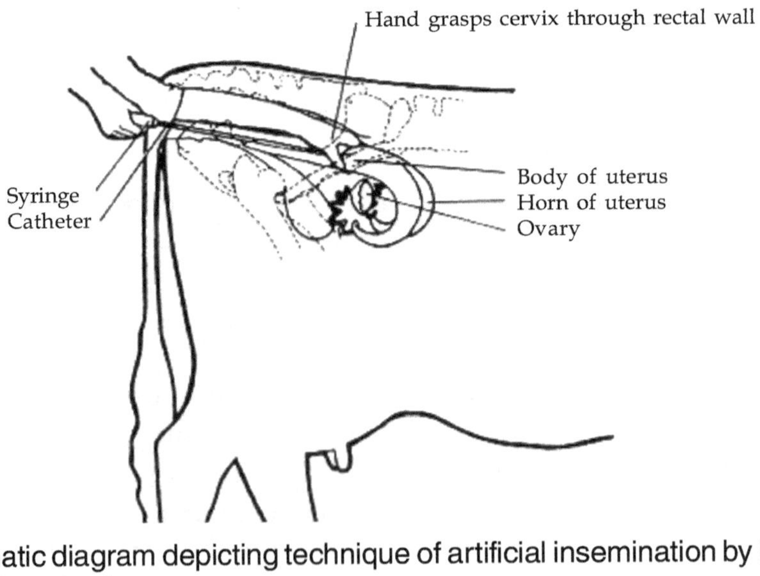

Schematic diagram depicting technique of artificial insemination by Recto Vaginal method

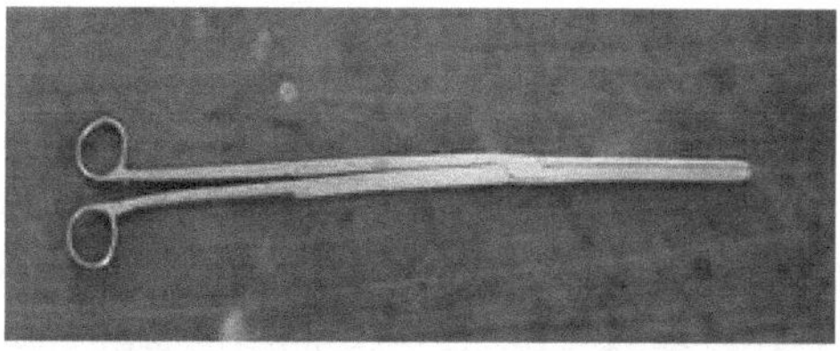

Forceps for retrieving straws from LN 2 cotainer

French type semen straw

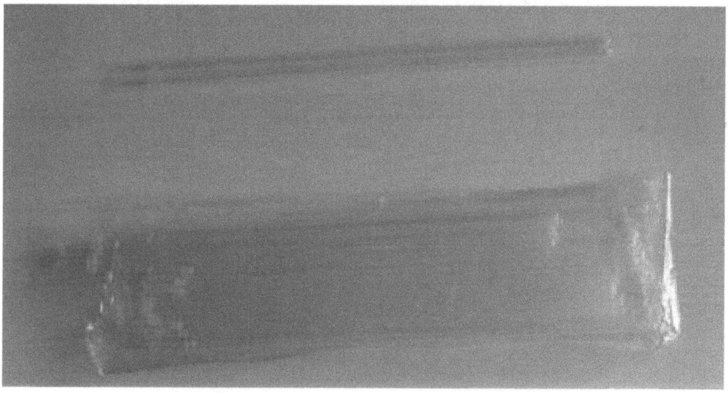

Artificial insemination sheath

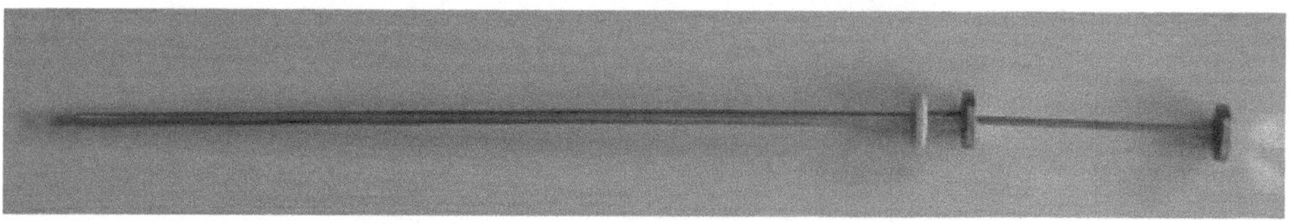

Loaded artificial insemination gun

e. When the bladder is extremely full, manipulate the clitoris to cause urination.

f. If the animal is straining hard or has ballooning of the rectum with air, do not struggle to inseminate the animal. Leave it for few minutes or may be hours and try again.

g. Try to maintain strict sanitary measures.

h. Be gentle to the animals.

Additional Pointers in Handling Semen for Insemination

1. Keep insemination equipment clean and dry at all times.
2. Check the accuracy of your thermometer used for recording temperature of water used for thawing occasionally.
3. Shake the straw as it is removed from the tank to remove any liquid nitrogen that may be retained in the cotton plug end of the straw.
4. Dry each straw of semen thoroughly. A small drop of water can be lethal to sperm.
5. Efficiency and technique of personnel performing AI should be evaluated on a regular basis. Proper AI technique should never be compromised for convenience.

Chapter 21

Field Handling and Upkeep of Liquid Nitrogen Containers, Semen Straws and Instruments Used in Artificial Insemination

21.1 Liquid Nitrogen Container/ Cryocan

To store liquid nitrogen a special type of container is used and is known as a LIQUID NITROGEN CONTAI NER OR CRYOCAN. Frozen semen doses are always preserved in liquid nitrogen which is an inert liquid having a boiling point of -196^{0}C and a freezing point of -212°C. Liquid Nitrogen Containers are doubled walled metal (Stainless Steel/Aluminium) vacuum vessels with a very efficient insulation system. Due to highly efficient insulation system the temperature inside the container can remain -196 °C as long as atleast two inches of Liquid Nitrogen is present in the container. It has got an inner chamber which is suspended in outer chamber through neck tube which is a non metal and bad conductor of heat. Walls of the inner chamber are coated with a high quality of insulation material. The space between the inner and outer chamber is also filled with high quality insulation material. Vacuum is created in between inner and outer chamber. Loss of vacuum will lead to the boiling of the liquid nitrogen gas and hence rapid loss of this gas from the inner chamber. The neck tube is the attachment between the inner and outer chambers and is non metallic and most delicate part of the whole Cryocan. The weight of the liquid nitrogen in the filled Cryocan has to be supported by neck tube. Therefore, sudden jerk and side to side movements of liquid nitrogen containers leads to mechanical damage of LN2 container. Appearance of frost line at the top of the container is sign of increased evaporation due to damage to the container. The main elements in maintaining the efficiency of the insulation system is the vacuum space which should be completely free from air molecules and the joint/welding between the inner and outer vessel at the point of neck tube. The liquid nitrogen containers are not as sturdy as they appear. They are very delicate and are susceptible to damage and injuries because of mishandling due to inadequate knowledge of their structure and function.

21.1.1 Points for Increasing the Efficiency of Liquid Nitrogen Containers

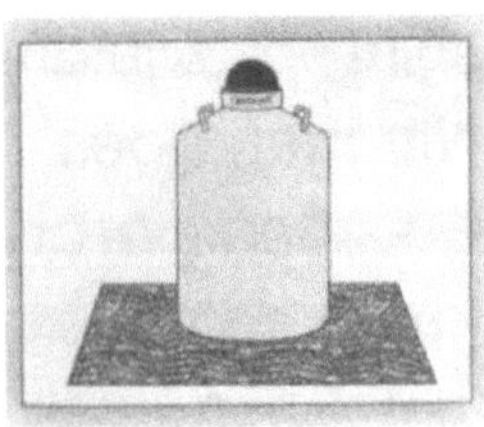

- Avoid a direct contact of liquid nitrogen containers with hard floor. Always keep LN2 container in a dry place on some rubber/ jute carpet or wooden plank. Contact with wet/ damp surface, hard cement floor and chemicals may cause corrosion and ultimately lead to vacuum loss. Moisture on the floor should also be avoided.

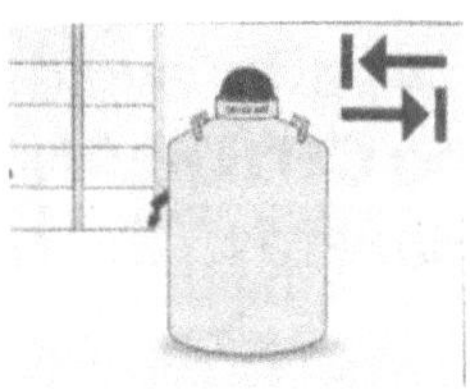

- Always keep the container vertically in a well ventilated room. Poorly ventilated room may lead to suffocation due to the decrease in atmospheric oxygen.
- Store the LN2 container in a cooler place to increase the life of the cryocan and increase the holding time of LN2. Avoid direct exposure to sun rays or hot air.

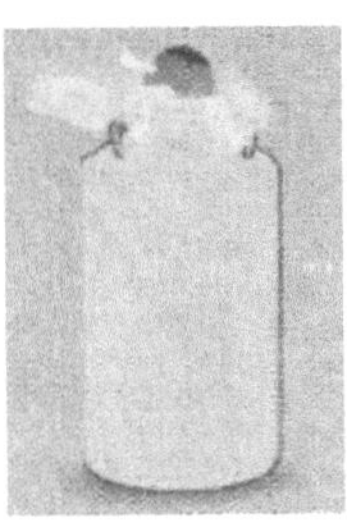

- Condensed moisture or frost formation on the outer shell of the container on the top of it and excessive evaporation of liquid nitrogen are indication of vacuum loss. If such condition arrives or is suspected, arrange for alternate cryocan immediately. Never play with vacuum knob of the liquid nitrogen containers as damage to it will lead to vacuum loss.
- Avoid all sorts of shocks like dropping/ rough handling. Always lift the container with both the hands and set it

down gently. Extensive shock may create a pendulum type vibration in the inner vessel and may lead to breakage at the neck tube junction. Do not expose the container to severe vibrations as this may damage the vacuum insulation system.

To avoid splashing of LN2, do not tilt/roll the LN2 container. This will also prevent dents/scratches which lead to vacuum loss. The liquid nitrogen container should always be shifted by platform base trolley.

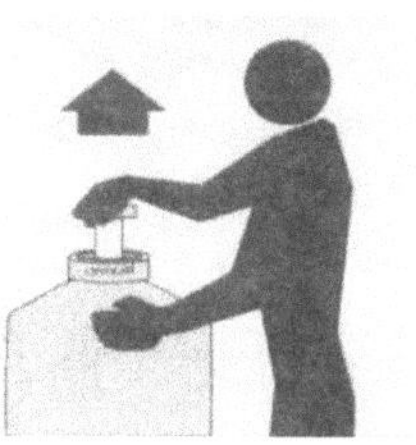

- The container should always be kept closed with neck plug to minimize the LN2 loss except the time of taking out the straws or filling of liquid nitrogen. Wipe the neck plug dry with cloth after usage in order to avoid ice formation and jamming of neck plug inside the Cryocan.
- Always slide out and put back the neck plug vertically while putting it on or off the cryocan. This will prevent possible damage of the neck plug.
- Never interchange the lid or canisters of the liquid nitrogen containers. Specific neck plug and canisters recommended for the individual model should always be used. Due to improper fitting of the neck plug there may be damage to the plug if it is too tight. Loosely fitted plug may lead to excessive loss of LN2 and frost formation leading to sticking of the plug and canisters to the neck tube. If canister is not specific for the model they may either not fit well or will not reach the boom of the inner chamber of the liquid nitrogen container.

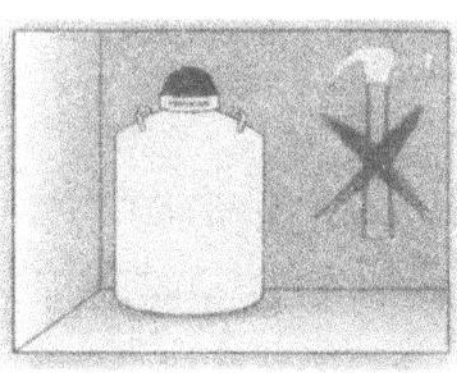

- Avoid scrapping, welding, brazing, drilling or punching on the wall of liquid nitrogen containers.

- Always fill the warm container slowly keeping clean the issuing gasses and slashing of liquid Nitrogen. Ensure that the Liquid Nitrogen does not come in contact with the Evacuation port. Opening and closing to insert and remove materials and moving the unit will also increase the evaporation rate.

- To prevent the Liquid Nitrogen Container from tipping, sliding and scraping during transportation it is preferable to use wooden crate/hard card board box with proper packing material. The Liquid Nitrogen Containers can also be secured with straps or with proper rubber band ring during transportation.

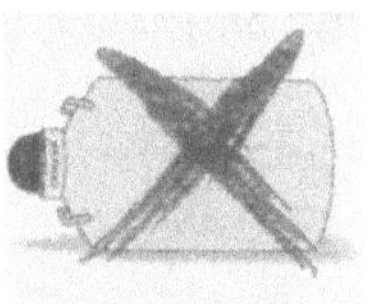

- Avoid tipping of the LN2 container and always keep it vertically. Do not let it lie on is side. This may lead to spillage and damage of the container and the materials stored in it.

- Use funnel to transfer the liquid nitrogen from one container to other. Never overfill the containers. Liquid withdrawal devices may be used when liquid nitrogen is to be transferred from heavier containers.

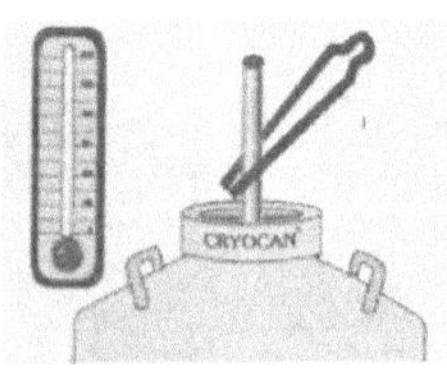

- Always use protective clothing while handing liquid nitrogen. Care should be taken to avoid direct contact of liquid nitrogen with body surface. Always use forceps to remove the straw from the LN2 container.

- Never fill liquid nitrogen in conainers which are in corrugated packing or in any other closed packing. This may lead to vacuum loss in the container through the evacuation port.

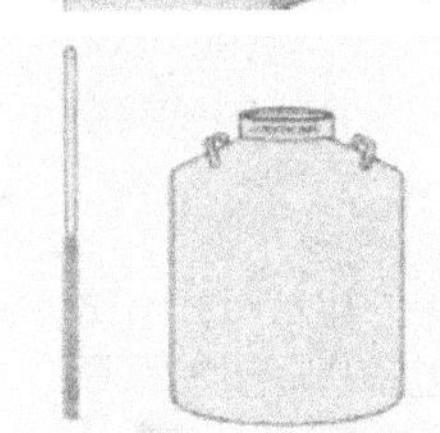

- The containers should be regularly checked for liquid nitrogen level and any increase in the rate of evaporation should be taken seriously and further investigated. Once a tank fails, nitrogen is lost very rapidly. Excessive evaporative loss is indicated by frost formation on the top of the container. Remember, even new tanks can have defects and fail. Plan to have an alternative semen tank available in case of a failure.

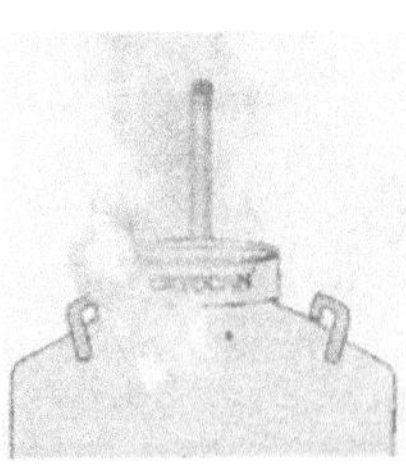

- Wooden or solid metal dipstick should be used for measuring the level of liquid nitrogen. Avoid using plastic sticks. Do not use hollow sticks as the liquid will spout from the bottom due to gasification and rapid expansion of liquid inside the tube.
- Do not stack LN2 containers one above the other.
- Store only the amount of semen needed for six months.

21.1.2 Dangers in Liquid Nitrogen Handling and its Precautions

1. Direct exposure of liquid nitrogen to the body surface may lead to tissue damage like frost bite as it is very cold (-196 ^{0}C). If it is accidentally splashed on the body surfaces use water to wash off the affected part.
2. Liquid nitrogen is non-toxic and non inflammable but continuous evaporation of liquid nitrogen in a poorly ventilated room causes decrease in atmospheric oxygen and may lead to suffocation. If a person becomes dizzy or unconscious while working with LN2 in a poorly ventilated room, move him to a well ventilated area immediately. Provide artificial respiration if the breathing has stopped. Oxygen may be given if required. Move the person at warm place and give rest. Call a physician immediately.
3. Introduction of warm objects in the liquid nitrogen may lead to boiling, splacing and gasification.
4. LN2 should not come in contact with rubber/plastic or any other material which cannot withstand a temperature of -196°C. This may cause serious damage to the material.
5. One should be extremely careful while transporting and handling LN2 especially in buses/trains or any gathering. Sudden evaporation may cause unnecessary terror/anxiety which may be responsible for accidents.

21.1.3 Determination of the Volume of Liquid Nitrogen in a Cryocan

1. Dip Stick Method

A dip stick can be used to measure the level of liquid nitrogen in a cryocan. A solid rod of low thermal conductivity such as wood, stainless steel or ebonite should be used as a dip stick. Do not use any hollow tube under any circumstances. Slowly insert the rod upto the bottom of the container vertically and keep it for 5 to 10 seconds. Take out the rod and wave it in the air. Measure the length of the frosted section formed after waving the dip stick in air. Determine the level of liquid nitrogen in the cryocan by comparing the frost level in the dip with the calibration chart of that particular model.

Do not do unnecessary and frequent measuring of liquid nitrogen levels as it may lead to excessive evaporative loss of liquid nitrogen. Always hang the dip stick hanged on the wall where liquid nitrogen containers are stored as keeping it loose any where may lead to breakage of the dipstick.

2. Weight Method

The volume of liquid nitrogen in a cryocan can be determined by measuring the weight of liquid nitrogen container in a sensitive balance. The difference of weight derived after subtracting the final weight from the initial weight gives the weight of LN2 lost. The amount of liquid nitrogen in the container can be determined by subtracting the weight of container containing LN2 from the weight of the empty cryocan. Multiplication of weight in kg by a factor 1.25 (specific gravity of liquid nitrogen) will give the weigh of LN2 in litres.

21.1.4 Testing Procedure for Newly Purchased Cryocans

For testing the Evaporation rate of LN2 at site for the newly purchased CRYOCANs, the following procedure may be followed:

1. The inner vessel of the new CRYOCANs at the Customer premises are at Room Temperature. Therefore when first charge of Liquid Nitrogen is filled in, the evaporation of Liquid Nitrogen will be excessive. This is so, because lot of Liquid Nitrogen will be consumed by the container for self cooling from temperature which can be as high as 40°C to Liquid Nitrogen temperature which is -196°C.
2. Do not get alarmed because of High Evaporation of First Filling. To allow the initial cooling of containers please fill Liquid Nitrogen upto 15-20% of specified volume during first filling.
3. Allow 24 Hours cooling time for the CRYOCAN.
4. During first filling frosting may appear in the cap. This frosting will take about two hours to disappear.
5. After 24 hours of initial cooling, fill up the container upto its full capacity taking care that liquid nitrogen is poured gently.
6. Normally the Container will take 10 days to achieve thermal balance hence the first reading should be noted accurately on Tenth Day.

Please Note

1. As per standard practice followed throughout the World the loss rate specified in our catalogue is calculated at 15°C temperature and 1.033 Kg.per sq.cm atmospheric pressure. Any deviation from this temperature and pressure may greatly influence the loss rate.

2. The loss rate can most accurately be determined by using an electronic weighing machine.
3. The dipstick method gives only representative loss rate which may be far from actual performance of the container.
4. When CRYOCAN is in actual use in the field, the loss rate will be much higher which corresponds to a field factor of 2 in Indian Field conditions. This means evaporation will be double than specified.
5. For details of testing procedure of Liquid Nitrogen Containers, the guidelines set by four National nominated Institutes may please be followed in addition to above.

21.2 AI gun

The AI gun consists of a barrel, a piston and a plastic 'O' lock. The AI gun should always be stored in a dry place in a plastic/ perpex container. It should be kept clean and hygienic. Avoid bending of AI gun and piston.

21.3 Plastic Sheath

The plastic sheaths should always be stored in a clean and dry place preferably in a plastic container. The plastic sheaths are commercially available in polyethene packet. The packet should not be opened unless the sheaths are to be used. When the sheaths are to be used, a small cut should be given at the side of the broad end of the sheath. After use the cut end should be folded and the sheath packet should be replaced in to the plastic container or an artery forceps may be applied on the cut end after folding it.

21.4 Important Points in Field Handling of Frozen Semen

21.4.1 During Semen Retrieval

Exposing semen to high temperature in the upper half of the neck tube during retrieval of semen dose for Artificial Insemination or transfer of semen dose from one container to another, can lead to severe decline in its quality on account of thermal damage. Thermal damage to spermatozoa is irreversible. Therefore, careful handling of frozen semen while its retrieval from frozen semen container should be done. The following points should be considered during semen retrieval from frozen semen container:

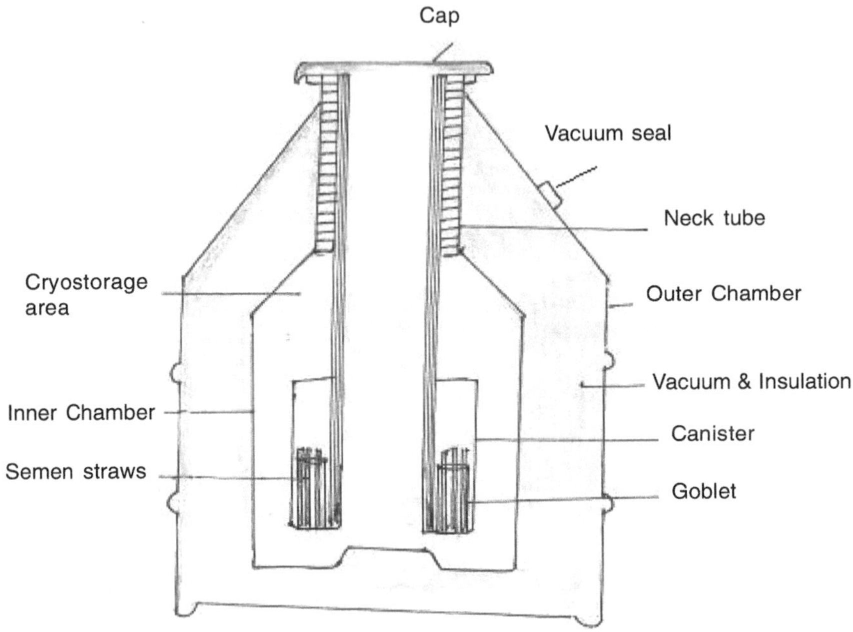

Inner construction of liquid nitrogen container

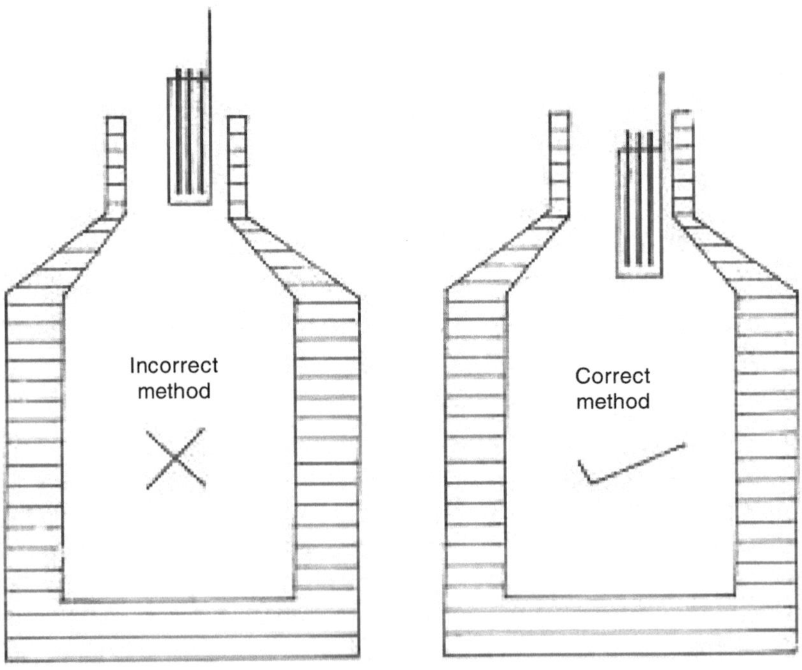

Correct and incorrect method of taking out straw from the LN 2 container

- Properly identify the canister containing the desired semen dose.
- Do not raise the canister containing the desired semen dose in the neck of the container above the frost line.
- Before holding the semen dose for retrieval pre cool the forceps in liquid nitrogen or its vapors for few seconds.
- After retrieving the desired frozen semen dose immediately lower the canister to the tank floor.
- The frozen semen should be removed within 10 seconds from the time the canister is raised in to position. If any time it takes more than 10 seconds to complete the task lower back the canister to the tank floor so as to enable it to dip in liquid nitrogen and then lift it again to complete the task.
- Once a semen dose is removed from the liquid nitrogen container it should never be returned back to the container.

21.4.2 During Insemination

Maintain the semen close to 35°C after thawing as many semen doses are have reduced quality between thawing and insemination. Therefore, precautions/ practices should be adopted at this vital step:

- While the semen dose is being thawed, warm the insemination gun by rubbing it briskly with a paper towel.
- After semen is thawed, wipe it dry with paper napkin/ towel and prevent it from rapid cooling.
- After thawing hold the straw at the laboratory end and shake it so that a air space occur at the laboratory end. This is done to prevent lose of semen while cutting the laboratory end of the straw. After shaking place the straw in the insemination gun so as to enable laboratory end to appear at the top.
- Cut the straw at right angle so as to achieve a good seal with the sheath and lose of semen in backflow while inseminating is not encountered.
- The cow should be inseminated within minutes of thawing (maximum 10-15 minutes).

21.4.3 Semen Transfer between Tanks

The While transferring semen from one container, the following points may be taken into account:

- Have tanks side by side before transferring semen. Fill the containers with liquid nitrogen.
- Use only appropriate canisters in Liquid nitrogen container. Never use wrong canister not compatible with the container.
- Transfer the semen dose within 3-5 seconds.
- Never touch the semen dose with bare hand. Proper handling and maintenance of frozen semen dose is a must.
- It is essential that the frozen semen dose be handled and thawed carefully to obtain optimum results.

Chapter 22

Cleaning and Sterilization of Equipments Used in Semen Processing and Artificial insemination

The equipments used in artificial insemination should be clean and sterilized. The use of unsterilized equipments in AI causes the spread of infection in the herd and infertility. Rinse the equipments with tap water immediately after use and then lukewarm detergent solution. The detergent used for washing the AI equipments should be non corrosive. Use of corrosive detergents will reduce half-life of the plastic and rubber wares. Wash thoroughly with tap water till the equipments become grease free and perfectly clean. Then rinse in distilled water at least thrice. Use of distilled water does not leave any salt residues on articles after drying.

Sterilization

Sterilization is the process of making the items free from any sort of living microorganisms and their spores. The equipments used in artificial insemination should be clean and sterilized. The use of unsterilized equipments in AI causes the spread of infection in the herd and infertility. Therefore, sterilization of consumables and glassware should be given top priority as the success of any frozen semen production and AI programme depends on it and, therefore, it should not be compromised at any cost. Rinse the equipments with tap water immediately after use and then lukewarm detergent solution. The detergent used for washing the AI equipments should be non corrosive. Use of corrosive equipments will reduce half life of the plastic and rubber wares. Wash thoroughly with tap water till the equipments become grease free and perfectly clean. Then rinse in distilled water at least thrice. Use of distilled water does not leave any salt residues on articles after drying.

Sterilization procedure depends on nature of material to be sterilized. The sterilization procedure of different materials used in production of frozen semen and AI is described below:

1. Glass Wares

Wash the glasswares thoroughly with tap water and dip the rubber wares in non-spermicidal neutral detergent for 30 minutes. Clean the rubber wares thoroughly with a nylon brush and rinse with running tap water followed by three rinses in double distilled water. Dry the glassware in hot air oven at 60°C. The open end of the glasswares should be covered with non absorbent cotton and sterilized in hot air oven at 160°C for one hour or 180°C for 30 minutes.

Sometimes the glassware does not look clean and becomes foggy. These should be treated with 2 parts of 0.5 % Potassium dichromate, 3 parts of Conc. H2SO4 and 25 parts of distilled water overnight before cleaning and sterilization as described above.

2. Rubber Wares

Wash the rubber wares thoroughly with tap water and dip the rubber wares in non-spermicidal neutral detergent for 30 minutes. Clean thoroughly with sponge brush and rinse with running tap water followed by three rinses in double distilled water. Plastic tips and tubings should be cleaned by applying jet of water with force by syringe. The rubber wares should be sterilized by Ethylene tri oxide gas sterilization. Relatively thermo resistant rubber wares should be sterilized by autoclaving at 3-4 psi for 10 minutes. The sterilized rubber wares should be kept in ultraviolet chamber.

3. Artificial Vagina (AV)

For best results the AV should be cleaned immediately after use in semen collection. Detach the cone from the AV and remove water from AV jacket before washing the AV. The cone and sponge should be cleaned with Soft sponge brush under running tap water and then soaked in warm solution of neutral non spermicidal detergent for around half an hour. Properly rinse in clean water followed by three times rinsing in double distilled water. After washing assemble the AV and autoclave at 5 psi pressure for 20 minutes. Keep the AV valve open during sterilization. Store the AV in incubator at 40-45°C to be used the next day for semen collection.

Cold washing method: Remove the cone from the AV and water from the AV jacket should be removed before washing. Put the AV along with the liner into a chlorhexadine disinfecting solution for 15 minutes. Similarly put the cone into this solution. Brush with neutral soap solution and warm water (not exceeding 60°C). Rinse three times with clear water followed rinsing with double distilled water. Rinse with small quantity of 70 % alcohol. Dry in a dust free area and then incubate the prepared AV at 44-46°C.

4. Collection Tube

Properly wash the collection tube under running tap water and then soaked in warm solution of neutral non spermicidal detergent for around half an hour. After this the collection tube is to be thoroughly brushed and then properly rinsed in clean water followed by three times rinsing in double distilled water. Dry the collection tubes in dryer or in incubator. Cover the mouth of the tubes with non-absorbable cotton/ aluminium foil and then sterilize in hot air oven at 160°C for 1 hour.

5. Distilled Water

The distilled water should be sterilized at 15 psi pressure for 20 minutes. Triple glass distilled water or Milli-Q water should be used for the preparation of dilutor.

6. Filter Papers

Wrap the filter papers used for separation of egg yolk in thick cloth and sterilize by autoclaving at 5 psi pressure for 20 minutes.

7. Buffer

The buffers used for preparation of extender should be sterilized at 5 psi pressure for 20 minutes.

8. Bacteriological Media

The bacteriological media used for colony counts should be sterilized at 15 psi pressure for 15 minutes.

9. AI Gun/ Other Metallic Instruments

AI gun and other metallic instruments should be autoclaved at 15 lbs pressure for 15 minutes.

10. Sheath, Straws and Working Surface

Ultraviolet radiation should be used for sterilization of sheath and straws.

11. Eggs/ Forceps for Breaking Eggs

The surface of the eggs used for preparing dilutor should be wiped with 70 % ethanol at the time of preparation of dilutor. The forceps used for breaking the eggs used should be wiped with 70 % ethanol.

Summary of Sterilization

Autoclave Items	**Pressure**	**Time (minutes)**
Artificial Vagina	5	20
Buffer	5	20
Plastic tips	5	20
Filter paper	5	20
Bull apron	5	20
Distilled Water	15	20
Bacteriological Media	15	15

Hot air oven

Item	**Temperature**	**Time (minutes)**
Glassware	160/ 180°C	60/30

Chapter 23

Maintenance of Records in Artificial Insemination System

Systematic recording is vital for smooth running of any artificial insemination enterprise. All the information in the center should be so arranged so that it can be readily be traced out and used for analysis of performance of the center. The records should be critically analyzed for selecting better animals / personnel and for improving reproductive management. The objective of record keeping is as follows.

1. To keep account of data regarding collection, evaluation and preservation of semen of individual bulls.
2. To keep account of Insemination and conception rate obtained from individual bulls.
3. To keep account of individual female inseminated and different operations carried out on each of them together with final result of every terminated service period.
4. To keep record of work carried out by each veterinarian or assistant employed at the artificial insemination center and result obtained by them.

The following format may be used for maintaining the records at artificial insemination center.

(A) Semen Collection Register

Department of Veterinary Andrology
Veterinary College, New Delhi

Serial No.	
1	Date and Time of Collection
2	Name & No. of the Bull
3	Breed
4	Reaction time in seconds
5	Volume
6	Colour
7	Consistency/ Density
8	pH
9	Mass activity (initial motility)
10	Individual motility
11	Sperm count per ml
12	Sperm count per ejaculate
13	% of dead and live sperms
14	Per cent abnormal sperm
	a. Head
	b. Midpiece
	c. Tail
15	Dilutor and Dilution rate
16	Metabolic test
17	Cold shock Resistance test
18	Initial of the officer
	Remarks if any

(B) Semen Record

Department of Veterinary Andrology
Veterinary College, New Delhi

1. Bull name/ number ....
2. Date of semen collection ....
3. Total ejaculate number ....
4. Yearly ejaculate number ....
5. Monthly ejaculate number ....

6. Time of semen collection ....
7. Reaction time (minutes) ....
8. Volume (ml) ....
9. Appearance ....
10. Consistency ....
11. Mass activity ....
12. Per cent motility ....
13. pH ....
14. Per cent Live spermatozoa ....
15. Sperm abnormalities
    a. Head ....
    b. Mid piece ....
    c. Tail ....
    d. Total ....
16. Sperm concentration .... (X 10^6 or million/ml)
17. Extender used ....
18. Dilution rate ....
19. Semen freezing
    a. Motility before straw filling ....
    b. Motility before freezing ....
    c. Post-thaw motility (0 hours) ....
    d. Post-thaw motility (48 hours)..

(C) Bull Performance Record

Department of Veterinary Andrology
Veterinary College, New Delhi

Name / No of the bull ....

Date of Birth/ Age ....

Breed .... Dam's milk yield ....

Sire's dam's milk yield ....

Dam's dam's milk yield ....

Daughter's milk yield ....

Sibling's milk yield ....

Date	Semen status Chilled/ Frozen	Motility	Cow/Buffalo Name/ No.	Owner	Number of AI (I,II,III or more)	Result (Pregnant/ Non Pregnant)	Signature of Inseminator	Remarks

Conception rate

1st Insemination ....

2nd Insemination ....

3rd Insemination ....

(D) Inseminator's Performance Record's

Department of Veterinary Andrology
Veterinary College, New Delhi

Month/ Year ....

Name of Inseminator ....

Name of the center ....

Date	Semen status			Cow/Buffalo		Result		Remarks
	Bull No	Chilled/ Frozen	Motility	Name/ No.	Owner	Pregnant	Non Pregnant	

Over all conception................................. %

FSC
www.fsc.org
MIX
Papier | Fördert gute Waldnutzung
FSC® C083411

Zeitfracht Medien GmbH
Ferdinand-Jühlke-Straße 7
99095 Erfurt, Deutschland
produktsicherheit@kolibri360.de